Polymer Properties at Room and Cryogenic Temperatures

THE INTERNATIONAL CRYOGENICS MONOGRAPH SERIES

General Editors	K. D. Timmerhaus, *Chemical Engineering Department, University of Colorado, Boulder, Colorado*
	Alan F. Clark, *National Institute of Standards and Technology Electricity, Gaithersburg, Maryland*
	Carlo Rizzuto, *Department of Physics University of Genoa, Genoa, Italy*
Founding Editor	K. Mendelssohn, F.R.S. *(deceased)*

Current volumes in this series

APPLIED SUPERCONDUCTIVITY, METALLURGY, AND PHYSICS OF TITANIUM ALLOYS • *E. W. Collings*
Volume 1: Fundamentals
Volume 2: Applications

CRYOCOOLERS • *G. Walker*
Part 1: Fundamentals
Part 2: Applications

CRYOGENIC PROCESS ENGINEERING
• *Klaus D. Timmerhaus and Thomas M. Flynn*

THE HALL EFFECT IN METALS AND ALLOYS • *C. M. Hurd*

HEAT TRANSFER AT LOW TEMPERATURE • *W. Frost*

HELIUM CRYOGENICS
• *Steven W. Van Sciver*

MODERN GAS-BASED TEMPERATURE AND PRESSURE MEASUREMENTS
• *Franco Pavese and Gianfranco Molinar*

POLYMER PROPERTIES AT ROOM AND CRYOGENIC TEMPERATURES • *Günther Hartwig*

STABILIZATION OF SUPERCONDUCTING MAGNETIC SYSTEMS • *V. A. Al'tov, V. B. Zenkevich, M. G. Kremlev, and V. V. Sychev*

SUPERCONDUCTING ELECTRON-OPTIC DEVICES • *I. Dietrich*

SUPERCONDUCTING MATERIALS
• *E. M. Savitskii, V. V. Baron, Yu V. Efimov, M. I. Bychkova, and L. F. Myzenkova*

Polymer Properties at Room and Cryogenic Temperatures

Günther Hartwig

Kernforschungszentrum Karlsruhe
Karlsruhe, Germany
and Universität Erlangen–Nürnberg
Erlangen, Germany

PLENUM PRESS • NEW YORK AND LONDON

Library of Congress Cataloging-in-Publication Data

Hartwig, Gunther.
　　Polymer properties at room and cryogenic temperatures / Günther Hartwig.
　　　　p.　　cm. -- (The international cryogenics monograph series)
　　Includes bibliographical references and index.
　　ISBN 0-306-44987-0
　　1. Polymers--Thermal properties.　2. Materials at low temperatures.　I. Title.　II. Series.
QC173.4.P65H37　1994
620.1'920421--dc20
　　　　　　　　　　　　　　　　　　　　　　　　　　　　　　94-43387
　　　　　　　　　　　　　　　　　　　　　　　　　　　　　　CIP

ISBN 0-306-44987-0

© 1994 Plenum Press, New York
A Division of Plenum Publishing Corporation
233 Spring Street, New York, N. Y. 10013

All rights reserved

No part of this book may be reproduced, stored in a retrieval system, or transmitted in any form or by any means, electronic, mechanical, photocopying, microfilming, recording, or otherwise, without written permission from the Publisher

Printed in the United States of America

*To my wife
Inge*

PREFACE

Most descriptions of polymers start at room temperature and end at the melting point. This textbook starts at very low temperatures and ends at room temperature. At low temperatures, may processes and relaxations are frozen which allows singular processes or separate relaxations to be studied. At room temperatures, or at the main glass transitions, many processes overlap and the properties are determined by relaxations. At low temperatures, there are temperature ranges with negligible influences by glass transitions. They can be used for investigating so-called basic properties which arise from principles of solid state physics. The chain structure of polymers, however, requires stringent modifications for establishing solid state physics of polymers. Several processes which are specific of polymers, occur only at low temperatures.

There are also technological aspects for considering polymers at low temperatures. More and more applications of polymeric materials in low-temperature technology appear. Some examples are thermal and electrical insulations, support elements for cryogenic devices, low-loss materials for high-frequency equipments. It is hoped that, in addition to the scientific part, a data collection in the appendix may help to apply polymers more intensively in low-temperature technology.

The author greatly appreciates the contributions by his coworkers of the Kernforschungszentrum Karlsruhe in measurement and discussion of many data presented in the textbook and its appendix. Fruitful disccussions with the colleagues Prof. H. Baur, Prof. S. Hunklinger, Prof. D. Munz and Prof. R. Schwarzl are greatly acknowledged. The author is very grateful to Dr. A. Clark for encouraging the work and reviewing the manuscript. Mrs. E. Nemes was responsible for typing; Mrs. U. Meyer-Paulus for the figures. Thanks are due to G. Christoph, Dr. C. Meingast and R. Hübner for proofreading. Last, but not least, I would like to thank my wife for her patience when sharing me with this book.

G. Hartwig

Karlsruhe, June 1994

CONTENTS

1. Introduction and General Polymer Features 1
2. Phonon Structure of Polymers 17
3. Specific Heat 47
4. Thermal Expansion and the Grueneisen Relation 67
5. Thermal Conductivity 97
6. Molecular Place Changes and Damping Spectra 117
7. Mechanical Deformation Behavior 139
8. Dielectric Properties and Their Correlations 173
9. Fracture Behavior of Polymers 187
10. Cryogenic Measuring Methods 219
11. Polymers as Matrix for Composites 241

Appendix: Data Base 251

Index 271

1.
INTRODUCTION AND GENERAL POLYMER FEATURES

INTRODUCTION AND GENERAL POLYMER FEATURES

Contents

1.	Introduction and general polymer features	3
1.1	Technological aspects of low-temperature investigations	3
1.2	Scientific aspects of low-temperature investigations	4
1.3	Polymer structure	6
1.4	Anisotropic and steric binding structures	8
1.5	Basic properties	10
1.6	Glass transitions and changes of place	11
1.7	Deformation behavior	14
1.8	Modes of vibration	15
1.9	References and general reading	15

Appendix

1A   Chemical composition of repetition units of polymers

1. INTRODUCTION AND GENERAL POLYMER FEATURES

1.1 TECHNOLOGICAL ASPECTS OF LOW-TEMPERATURE INVESTIGATIONS

Important areas of application of polymeric materials at low temperatures are:

- Superconductivity (fusion technology, cryogenic electronics),
- Space technology (cooling of equipment, cryogenic wind tunnels),
- Storage and transportation of liquid gases (e.g., hydrogen technology).

Polymers are commonly used in electrical insulations and as matrices for composites, especially for fiber composites. For most cryogenic applications polymer based materials are required, in addition, for thermal insulation.

Nonmetallic helium cryostats are necessary in several applications, e.g., for SQUID detectors. For support structures and cryostats of pulsed superconducting magnets nonmetallic materials are required when heating by eddy currents is to be avoided.

Hydrogen technology is rapidly progressing. Liquid hydrogen vessels for vehicles will be built with lightweight materials of high specific strength and fatigue resistance and low thermal conductivity. Fiber composites with polymers are prime candidates. Wind tunnels yield much higher Reynolds numbers when operated at cryogenic temperatures. The favored materials for the paddles are carbon fiber composites which have a high specific strength and a high fatigue endurance limit.

Support elements which link cryogenic devices to room temperature environment require a high mechanical strength and a particularly high thermal insulating power. Carbon- and ceramic fiber composites exhibit a very low thermal conductivity at low temperatures.

Problems arise for most designs of cryogenic support structures from the thermal expansion, especially if different materials are involved. Special fiber composites have been developed which exhibit a negative thermal expansion coefficient (nearly that of steel, but with a negative sign [1.1]). They can be used as compensating elements in cryogenic devices. Thermal stresses caused by the mismatch of thermal expansion are a general problem in cooling. They become serious when they preload polymers, which are very brittle at low temperatures. Consequently, ductile polymers were searched for. Several thermoplastic polymers have been found that exhibit some ductility (e.g., ultimate strain $\approx 4\%$ at 4.2K for PEEK). However, these polymers have the disadvantage of poor creep resistance even at low temperatures.

The knowledge of polymer properties is necessary to calculate the properties of composites. Some of them are strongly influenced by the specific matrix properties. Several low-temperature properties of polymers are rather insensitive to chemical compositions, but thermal conductivity, dielectric damping, fracture, and fatigue behavior depend very much on the type of polymer, and a proper selection has to be made. The knowledge of polymer properties helps in selecting the proper matrix materials. This topic will be briefly discussed in Chapter 11.

1.2 SCIENTIFIC ASPECTS OF LOW-TEMPERATURE INVESTIGATIONS

The complicated inhomogeneous microstructures of polymers, the different binding forces within and between chains, the great variety of chemical compositions, and the possibility of place changes of molecular groups makes a general treatment of polymer properties very difficult. This is especially true at higher temperatures. At main glass transitions many relaxation processes overlap, and single relaxations cannot be analyzed. At lower temperatures, however, binding forces, single relaxations, and several rules or individual processes can be studied separately. This will be discussed in the following subsections.

1.2.1 Glass Transitions of Individual Molecular Groups

Most properties of amorphous polymers are influenced by several weak glass transitions well below the primary glass transition temperature. They arise from unfreezing of the mobilities of specific molecular groups. (The nomenclature of secondary and tertiary glass transitions is related to the mobilities of small chain segments or side groups, respectively; see Section 6.2). The analysis of those individual mobilities is a domain of low-temperature investigations. Glass transitions have been found down to 30K. However, for most of the polymers they do not occur below 120K.

1.2.2 Tunneling Processes

Tunneling processes in amorphous polymers become dominant near or below 10K. They influence several properties and are a specific topic of investigations at very low temperatures.

1.2.3 Basic Properties

There is a range of temperatures, where the influence on properties by tunneling processes are no longer and those by glass transitions are not yet effective. It extends at least from 10K to 40K; in fact, for most polymers the range is much larger. This allows the "basic properties", which are closely connected to solid state physics, to be investigated (see Section 1.5). Several rules and principles of solid state physics can be applied to polymers at low temperatures. A great advantage is that several low-temperature properties are rather independent of chemical compositions; they depend, however, on parameters such as crystalline content or cross-link density. Dielectric properties, in addition, are functions of the dipole moment of polymers (see Section 8.5). A schematic diagram in Figure 1.1 gives a survey of the properties and dominant processes in certain temperature ranges.

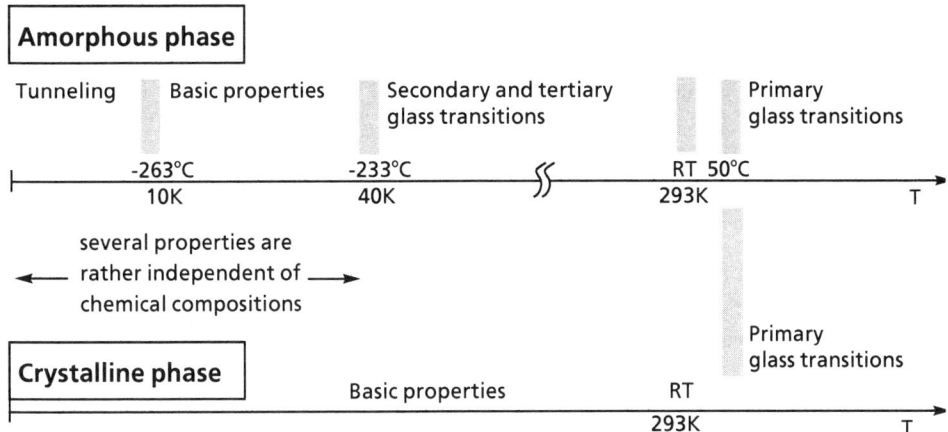

Fig. 1.1. Schematic diagram of properties and influencing processes versus temperature.

It is obvious that amorphous polymers are subjected to several specific processes. The great variation of polymer properties at elevated temperatures is not so much influenced by the chemical composition, but arises mainly from relaxation processes. An ideal crystalline polymer, by contrast, would not undergo secondary relaxations, and "basic properties" would be dominant until the primary glass transition. The properties of semicrystalline polymers are determined by the crystalline phase and even more by the amorphous phase.

1.2.4 Inter- and Intrachain Interactions

The polymer properties are determined by:

- Interchain interactions (between chains),
- Intrachain interactions (within a chain).

Interchain Interactions. At low temperatures, polymeric chains get frozen, and thermal vibrations and mechanical deformations (e.g., moduli) are controlled mainly by the weak interchain binding forces (Van der Waals forces). Thermal low-temperature properties are dominated by long-wave vibrations, which are able to propagate between chains (interchain vibrations). Elastic properties are determined mainly by the interchain force constant given by the Van der Waals forces. Interchain binding forces can be analyzed at low temperatures without regard to viscoelastic intrachain effects.

Intrachain Interactions. At higher temperatures molecular chains become viscoelastic, and deformations (e.g., moduli) are controlled mainly by intrachain forces. At elevated temperatures thermal properties are dominated by short-wave vibrations, which propagate along the chains only (intrachain vibrations). The relative contributions to the properties by inter- and intrachain interactions are shown schematically in Figure 1.2. The study of the extent to which Van der Waals potentials influence physical properties is a topic of many low-temperature investigations.

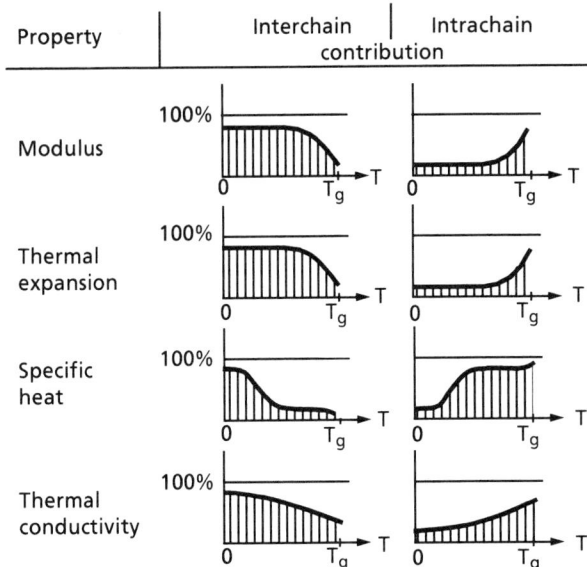

Fig. 1.2. Schematic representation of the relative influence of inter- and intrachain interactions on important polymer properties versus temperature. (T_g is the main glass transition temperature).

1.2.5 Boundary Effects

At low temperatures boundary scattering of phonons reduces the thermal conductivity drastically. This is a typical topic of low-temperature investigations. Boundaries occur in filled polymers or in semicrystalline polymers.

1.3. POLYMER STRUCTURE

Polymers are characterized by their chain structure, the arrangements of chains, and the mixture of different kinds of chains or phases.

1.3.1 Classification Related to the Structure and Arrangement of Chains

The **chain structure** is given by the following characteristics:
- Chemical compositions of the monomers (see Appendix 1A);
- Configurations (relative arrangement of monomers within a chain; e.g., isotactic, atactic, syndiotactic);
- Conformations (geometrical arrangement; relative rotations of chain segments; e.g., trans-, gauge conformations).

Homopolymers are polymerized from equal monomers, **copolymers** from different ones.

The **chain arrangements** are given by the following topographies:
- linear chains (e.g., HDPE, PEEK);[1]
- branched chains (e.g., LDPE); they are a group between linear and cross-linked polymers.
- chains with side groups (e.g., PMMA, PS);
- cross-linked network (e.g., epoxy resins, phenolic resins).

---
[1] Abbreviations are representative of the chemical compositions and are listed in Appendix 1A.

1.3.2 Classification Related to the Conditions of Processing

A widely used classification of polymers reflects their types of processing.

Thermoplastic Polymers. The polymer chains are weakly bonded by Van der Waals forces. They are hot meltable and soluble which favors processing by extrusion and casting.

Amorphous Thermoplastic Polymers. They consist of entangled chains.

Semicrystalline Polymers. Small crystallites are embedded in the amorphous phase. The crystallites have different structures (e.g., helical or meander- and folded lamellar conformations, see Figure 1.3b; superstructures, such as spherulites, are known, e.g., for PE). Many polymers are able to form both amorphous and crystalline phases. Their relative fractions in semicrystalline polymers depend on several parameters, especially on the cooling rate when they are prepared.

Blends are mixtures of different kinds of molecular chains (e.g., SAN, styrol-acrylonitrile).

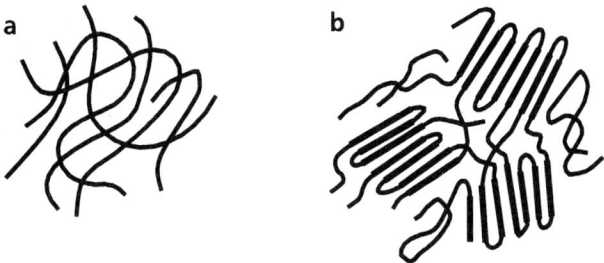

Fig. 1.3. Structures of thermoplastic polymers, a) amorphous; b) semicrystalline (folded lamellar crystallites).

Liquid Crystalline Polymers (LCPs). LCPs are polymerized from short but very stiff chain segments, which can be oriented, e.g., by shear flow during extrusion. They feature self-orientation in the mesomorphic phase (between liquid and solid phases). Depending on the production process, a more or less oriented chain structure can be achieved. The properties are then anisotropic. Pure LCPs have a linear chain structure but are not meltable like thermoplastic polymers. They are meltable only as copolymers with thermoplastic "spacers."

Duroplastic Polymers (Thermosets). Duroplastic polymers are two-component systems polymerized from araldites and hardeners. During curing they become cross-linked (Fig. 1.4a). For instance, three-functional epoxy groups can cause chemical cross-linking (Fig. 1.4b). After curing they are not meltable and hardly soluble. Well-known and widely used thermosets are epoxy resins, whose properties can be varied by the cross-link density and the type of components. Highly cross-linked epoxy resins are rigid and quite brittle.

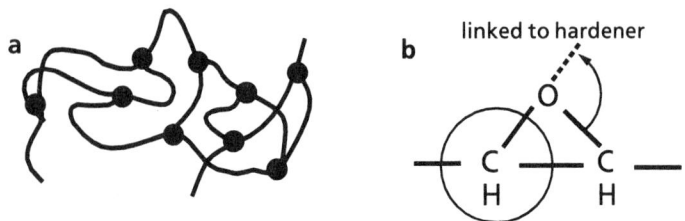

Fig. 1.4. a) Cross-linked epoxy resins; b) three functional epoxy group.

1.3.3 Models of Polymer Structures

The morphology of polymers is rather complex. The entangled conformation of amorphous polymers is difficult to treat, and approximations are necessary. The zigzag or helical chain structure is a further complication. Aligned polymers are widely used as models. Figure 1.5a shows a model of aligned zigzag chains. The dashed lines represent the Van der Waals forces. In most cases the rough approximation of aligned linearized chains is applied. This model shown in Figure 1.5b will be applied in Chapters 3 and 4 to calculate inter- and intrachain vibrations. It can be applied in microdomains even to entangled polymers to calculate thermal properties. A curved coordinate system can be assumed which follows the mean chain orientation over short distances. For instance, longitudinal and transverse vibrations can be treated rather successfully in this way.

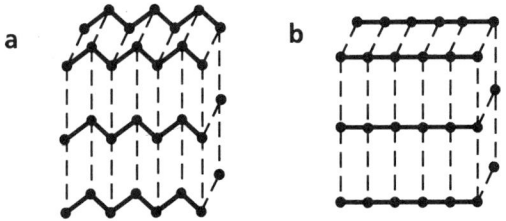

Fig. 1.5. a) model of aligned zigzag chains; b) model of linearized chains.

1.4 ANISOTROPIC AND STERIC BINDING STRUCTURES

The large anisotropy of binding forces is very pronounced in aligned polymers. Strong covalent forces act in the direction of polymeric chains, while perpendicular to it molecules are bound by Van der Waals forces, which are weaker by about two orders of magnitude.

1.4.1 Covalent Forces

Covalent forces arise from quantum-mechanical exchange interactions of electrons shared by two atoms. They exhibit a spatial distribution and form a binding structure with valence angles. The zigzag structure of polymer backbones is an example.

1.4.2 Van der Waals Forces

Van der Waals forces consist of contributions from several binding forces:

- Dispersion forces.
- Permanent (or induced) dipole forces.

The **dispersion forces** are calculated by the second-order perturbation theory. They are isotropic and independent of the permanent dipole moments of the molecular units. They exist even for nonpolar molecules.

The **dipole forces** arise from classical interactions between permanent dipoles and are usually incorporated into the Van der Waals force term. When permanent dipole moments are involved, Van der Waals forces also exhibit some spatial distribution. Dipole forces are smaller than dispersion forces.

The potentials φ of all these forces are equally dependent on the relative distance r between interacting particles:

$$-\phi = \frac{a_0}{r^6} - \frac{b_0}{r^{12}}. \tag{1.1}$$

The first term describes the attractive energy while the last term has been found empirically for the repulsive energy. The coefficients a_0 and b_0 are very different for covalent and Van der Waals bonds. As seen from Figure 1.6, a potential minimum always exists which characterizes the distance atoms ($r_c \approx 0.12$nm) or molecular chains ($r_w \approx 0.45$nm) at absolute zero temperature.

Since polymers normally freeze in a nonequilibrium state, these minima are relative ones and do not represent the minimum free energy. The potentials are nonharmonic. Their asymmetry controls thermal properties such as thermal expansion and thermal conductivity (see Chapters 4 and 5). A harmonic description is valid only at low temperatures. The vicinity of the minimum can be approximated by a circular segment, whose curvature can be determined from low-temperature measurements (see Chapter 6). The origin of those binding forces are central forces, which are dependent on the distance between interacting particles. They are responsible for stretching mode vibrations in thermodynamics and for ideal-elastic properties in mechanics.

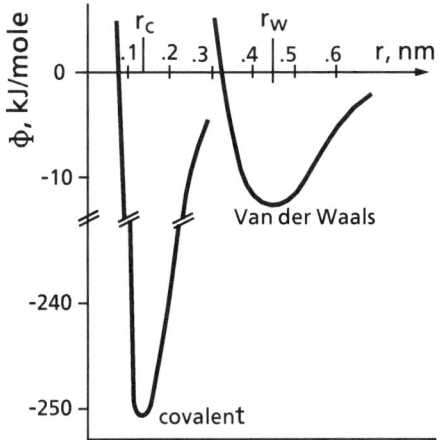

Fig. 1.6. Covalent and Van der Waals binding energies as a function of the distance r.
r_c : atomic distance,
r_w : chain distance.

Covalent binding structures, in addition, imply angular components. The backbones of polymeric molecules are formed mainly by carbon atoms, which show up as a tetrahedral binding structure. Valence angle deformations control bending mode vibrations perpendicular to the chain direction (see Section 2.3).

1.4.3 Rotational Binding Potentials

These potentials control rotations of chain segments or side groups. The rotation around a C−C bond is *a priori* free. Potential barriers against rotations arise from neighboring chains and from Van der Waals interactions of side substituents (see Fig. 1.8 and Section 6.3). As already mentioned, they determine the conformation of polymer chains.

1.4.4 Hydrogen Bond

Another type of binding forces whose strength is between that of covalent and Van der Waals binding forces is the so-called hydrogen bond. It is named hydrogen bond since a hydrogen atom is shared by two neighboring segments which contain strongly electronegative atoms, such as F, O, or N. These forces cause an additional binding between molecular chains of special polymers (e.g., aramid molecules of Kevlar) [1.2].

1.5 BASIC PROPERTIES

The great differences of polymer properties at elevated temperatures mainly arise from glass transitions, so that it is difficult to describe them by common solid state physics. At low temperatures, however, the situation is more uniform and several mechanical and thermal properties are not strongly dependent on the chemical compositions of polymers. One reason is that at low temperatures nearly no glass transitions occur. This leads to the introduction of the so-called basic properties, which comprise only those components which are not influenced by the great variety of glass transitions. They are more closely related to solid state physics. As already mentioned, they can be studied at low temperatures. At higher temperatures there is a superposition of basic properties and those which result from glass transitions.

1.5.1 Influencing Parameters

Experiments revealed that polymer properties at low temperatures are not strongly dependent on their chemical compositions. (The influences of molar mass, moisture or water content are less serious below ~100K.) There are, however, several parameters of conformation, which control the basic properties:

- Crystalline content of semicrystalline polymers.
- Cross-link density of duroplastic polymers [1.3].

The arrows in Figure 1.7 show the tendency of the influences on properties. The thermal conductivity of cross-linked polymers is influenced only at low temperatures, at which the phonon wavelength is on the order of the cross-link distance. For semicrystalline polymers below 20K boundary effects, which reduce the thermal conductivity, become dominant (see Section 5.4).

INTRODUCTION AND GENERAL POLYMER FEATURES

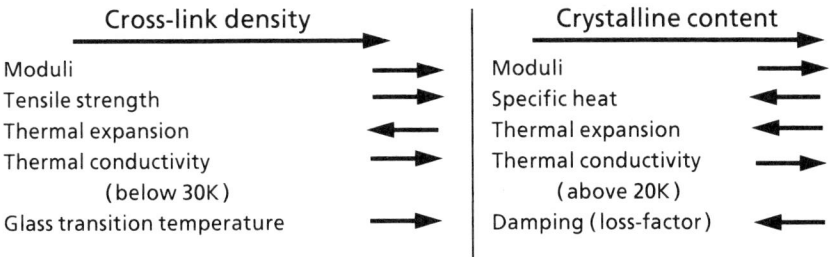

Fig. 1.7. Influence of cross-link density and crystalline content on polymer properties.

1.5.2 Asymptotic Low-Temperature Behavior

At low temperatures (T < 10K) several properties approach a narrow band of values. This is true for the specific heat, the thermal expansion, the mechanical damping, the elastic moduli, and the dielectric permittivity. However, the band of values depends on the crystalline content or on cross-linking. Dielectric damping, in addition, depends on the permanent dipole moment.

There are several reasons for this behavior. As seen in Figure 1.9, several properties of amorphous polymers at low temperatures are controlled by tunneling processes, which are rather insensitive to materials. Mechanical damping, e.g., at very low temperatures, is determined by tunneling processes, especially in the plateau region. Elastic properties are determined mainly by Van der Waals forces, which are quite similar for most amorphous polymers (semicrystalline modifications show higher moduli due to their tighter packing). At low temperatures, the dielectric permittivity is determined mainly by the electron components, which are rather insensitive to the chemical composition of polymers. Thermal properties are not very material-sensitive since at low temperatures only long-wave phonons are activated which do not resolve the material structure.

1.6 GLASS TRANSITIONS AND CHANGES OF PLACE

1.6.1 Types of Place Changes

A glass transition is characterized by unfreezing of molecular mobilities upon warming. This occurs in a narrow temperature range. For primary glass transitions molecular chains become fully flexible and many relaxation processes overlap. The glass transition, however, can be characterized by a (time-dependent) glass transition temperature T_g.

At low and medium temperatures only small individual molecular parts become mobile and take part in a weak glass transition, which again is characterized by an individual (time-dependent) glass transition temperature:

T_{gs}: secondary glass transition (motions within a chain);

T_{gt}: tertiary glass transition (side group motions).

The physical cause of glass transitions of polymers are place changes of molecular groups. The loosely packed structure of amorphous polymers allows certain segments or side groups the possibility of two positions, which are characterized by a double well potential. Place changes can be activated by mechanical or dielectric loading. Two types of place change occur (Fig. 1.8):

- Tunneling (dominant well below 10K) [1.4].
- Potential barrier jumping (T ≥ 40K).

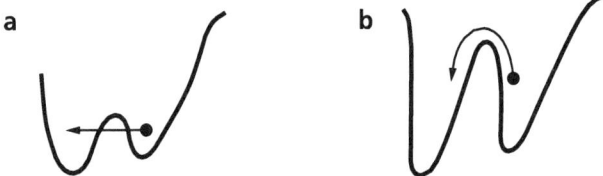

Fig. 1.8. Double well potentials with different types of place change:
a) tunneling; b) potential barrier jumping.

Tunneling Processes. It is well known in quantum-mechanics that wave functions of particles penetrate barriers and give a finite probability of particles to exist on the other side of the barrier. This process violates the principle of conservation of momentum at a certain distance (barrier thickness). The probability of tunneling transitions is quite small. They are dominant only if thermal processes of barrier jumping die out at very low temperatures.

Barrier Jumping Processes. At elevated temperatures particles within a certain statistical distribution have enough thermal energy to overcome potential barriers. At increased temperatures higher barriers can be jumped.

1.6.2 Relaxation Times

Place changes by tunneling or jumping processes are statistical processes which take some time, the so-called relaxation time. It can be determined by application of an external load which disturbs the state of thermal equilibrium. The relaxation time is necessary to restore the thermal equilibrium. The relaxation time, therefore, depends on the temperature and becomes longer at lower temperatures. At very low temperatures, the relaxation time becomes shorter again since tunneling transitions become dominant. They open the door to a great variety of unexpected physical properties. Relaxation times of tunneling processes have a rather broad and smooth distribution. By contrast, at secondary or tertiary glass transitions the barrier jumping processes are due to specific molecular motions all of which exhibit rather distinct relaxation times. They can be used to calculate the potential barriers (see Section 6.3).

Quasi-periodic transitions between potential minima can be considered to constitute vibrations and, consequently, they influence most of the thermal properties. The relaxation time is a measure of the mean frequency of these transitions. In addition, at each glass transition the free volume is changed and several properties are influenced.

1.6.3 Addition of Properties

The properties of the amorphous phase of polymers are drastically influenced by place change processes. Most properties are superimpositions of basic properties and those arising from relaxation processes by tunneling or barrier jumping. A schematic diagram is shown in Figure 1.9. The dashed lines represent the basic properties without relaxation effects. The solid lines mark the influence of relaxations.

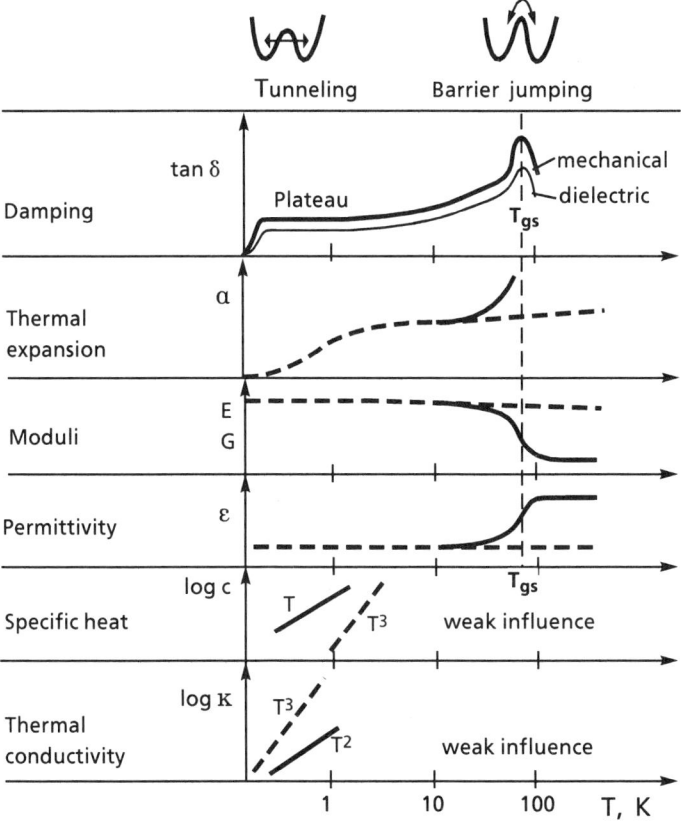

Fig. 1.9. Effect of place changes on several physical properties. A secondary glass transition temperature is marked by T_{gs}.

The properties of semicrystalline polymers are determined by the contributions from their amorphous and crystalline components. Crystalline domains of polymers are sometimes more tightly packed, thus leaving no free volume for two or more positions of a molecular segment. They do not show secondary or tertiary glass transitions.

Tunneling processes determine mechanical and dielectric damping below 10K and specific heat and thermal conductivity below 1K. The temperature dependence of thermal conductivity in Figure 1.9 is valid only for a completely amorphous polymer (solid line) or a fully crystalline polymer with basic properties (dashed line). For semicrystalline polymers the temperature dependence is different since boundary effects become dominant at low temperatures.

At elevated temperatures the influence of relaxations by barrier jumping becomes dominant. In this representation a secondary glass transition is assumed to take place at a temperature T_{gs} at which mechanical and dielectric damping maxima exist and the elastic moduli, permittivity, and thermal expansion are drastically changed. The influence on the specific heat and thermal conductivity is weak and smeared out.

1.7 DEFORMATION BEHAVIOR

The deformation behavior of polymers is characterized by different processes which depend on the distribution of the binding potentials. A schematic diagram is shown in Figure 1.10. The following types of deformations are characteristic of polymers:

- **Elastic deformations** result from deformations of binding electrons (electron orbitals) without place changes. They are:

    reversible, nondissipative, and time and temperature independent.

- **Viscoelastic processes** involve place changes of molecular units and are:

    reversible, dissipative, and dependent on time and temperature.

- **Viscous processes** involve place changes over several potential barriers. They occur if the thermal energy plus deformation energy are comparable to the potential barriers. They are:

    irreversible, dissipative, and time and temperature dependent.

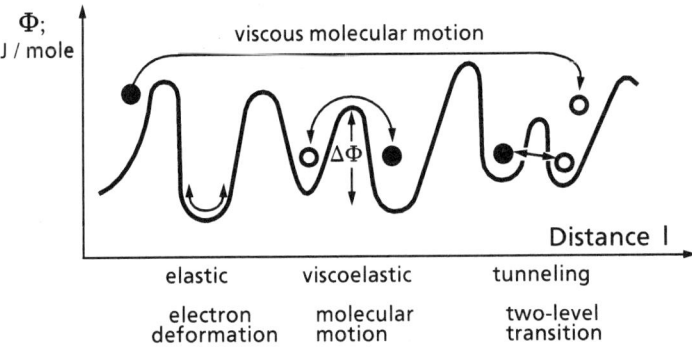

Fig. 1.10. Potential distribution and deformation processes of polymers.

Tunneling processes are represented in Figure 1.10 for the sake of completeness. Their influence on the deformation behavior of polymers is small. Viscoelastic and viscous deformation processes are characteristic of polymers. At low temperatures viscous processes are nearly negligible. At moderate temperatures, however, large mechanical stresses may lead to some irreversible deformation and orientation for thermoplastic polymers.

1.8 MODES OF VIBRATION

The chain structure of polymers and their large free volume give rise to different and specific modes of vibration.

Stretching (or longitudinal) vibrations within and between chains are driven by central forces of the binding potentials. The relevant force constants can be determined from the moduli along and perpendicular to the chain direction (see Section 2.3).

Bending (or transverse) vibrations are characteristic of polymers. They are driven by the intrinsic bending stiffness of polymer chains, which arises from the stiffness against valence angle deformations. It can be shown that vibrations transverse to the chain axis are controlled mainly by the force constant of bending stiffness.

Several thermal properties are influenced by these modes of vibration. This is especially true for the specific heat and thermal expansion. Bending (or transverse) vibrations cause a temperature dependence of specific heat, which is different from that of longitudinal vibrations (see Section 3.3). Thermal expansion resulting from transverse vibrations is very different from that of longitudinal vibrations. In most cases, a negative coefficient of thermal expansion has been found when transverse vibrations are dominant (see Section 4.3).

These types of vibration can be described by phonons (see Chapter 2). Their density distribution depends on the mode of vibration. Low-frequency (acoustic) phonons in many cases are sufficient to describe thermal properties at low temperatures.

There are other types of vibration, which are indirectly associated with phonons.

Tunneling vibrations are excited by phonons that are absorbed by tunneling systems (see Section 8.6). As already mentioned, they effect periodic tunneling transitions of very small molecular units. Those vibrations are localized and dominant at very low temperatures. Below 1K they influence several thermal properties.

Optical mode phonons are induced by absorption of infrared photons. They influence the dielectric properties. Their influence on thermal low-temperature properties is small.

1.9 REFERENCES

1.1 Hartwig, G; Status and Future of Fibre Composites; Adv. in Cryog. Engineering, **Vol. 40**; p.961; Plenum Press; New York; (1994).

1.2 Hartwig, G; Overview of Advanced Fibre Composites; in Cryogenics **28**, p.216; (1988).

1.3 Fischer, M.: Properties and Failure of Polymers with Tailored Distances between Crosslinks; Adv. Polymer Science, **100**, pp.315 to 333; (1992).

1.4 Hunklinger, S. and A.K. Raychandhuri; Progr. in Low-Temp. Phys.**IX**; p.256, Ed.: D.F. Brewer; Elsevier Sci. Publ.; (1986).

GENERAL READING

1. Polymeric Materials; ASM : American Society for Metals; Metals Park, Ohio 44073; (1974)
2. Kittel, C.E.; Introduction to Solid State Physics; John Wiley and Sons, New York (1973).
3. Mc Crum, N.G., B.E. Read and G. Williams; Anelastic and Dielectric Effects in Polymeric Solids; John Wiley and Sons, New York.
4. Perepechko, J.; Low-Temperature Properties of Polymers; Pergamon Press, MIR Publishers, Moscow; (1980).
5. Hedvig, P.; Dielectric Spectroscopy of Polymers; Adam Hilger LTD, Bristol (1977).
6. Krevelen, D.W. and P.J. Hoftyzer; Properties of Polymers (correlation with chemical structure); Elsevier Publishing Company (1972).
7. Hartwig, G.; in: Tieftemperaturtechnologie; pp.139-183; Ed.: F.X.Eder; VDI-Verlag (1981).
8. Encyclopedia of Polymer Science and Engineering; **Vol. 4**; Second Ed. (1986) and Concise Encyclopedia of Polymer Science and Engineering; John Wiley and Sons; New York; (1990)
9. Conference Proceedings : Nonmetallic Materials and Composites at Low Temperatures, **Vol. 1** (1978), **Vol. 2** (1980), **Vol. 3** (1984); Eds. : G. Hartwig and D. Evans, Plenum Press.
10. Cryogenics : Special issues (Conference Proceedings), **Vol. 28** (1988) and **Vol. 31** (1991).
11. Schwarzl, R.F. ; Polymer Mechanik; Springer Press; Heidelberg (1990).
12. Hawards, R.N.; The physics of glassy polymers; Applied Sci. Publishers Ltd., London, (1973).
13. Hunklinger, S.; in: Phonon scattering in condensed; Solid State Science **51**; Eds.: W. Eisenmenger; K. Laßmann and S. Döttinger; Springer Press; Berlin, (1984).
14. Reissland, J.A.; The Physics of Phonons; John Wiley and Sons; New York.

PHONON STRUCTURE OF POLYMERS

Contents

2.	Phonon structure of polymers	19
2.1	General introduction	19
2.2	Modes of vibration	23
2.3	Dispersion relations and phonon velocity	25
2.4	Density of states	35
2.5	Probability of states	44
2.6	Summary	45
2.7	References	46

2 PHONON STRUCTURE OF POLYMERS

2.1 GENERAL INTRODUCTION

Historically, interactions in physics have been described by exchanges of particles. This idea was transposed to phenomena such as light, sound, magnetism or polarization. Photons, phonons, magnetons or polarons are phenomenological particles provided with specific properties which allow interactions with themselves and with solids to be described. All of these "particles" are quantum mechanical in their nature and share some basic features:

- dual corpuscular and wave character,
- quantized exchange of energy, momentum or spin.

Thermal or acoustic vibrations of solids are usually described by phonons. The term phonon comprises frequency, wavelength, energy, momentum, direction of propagation, polarization and type (mode) of vibrations. There are several correlations among these parameters, which will be briefly described in the following sections.

The phonon energy E is calculated by the angular frequency ω and Planck's constant $h \equiv 2\pi \hbar$.

$$E = \hbar \omega \quad (2.1)$$

Another parameter, analogous to the classical momentum $\vec{p}$, is derived from the wave vector $\vec{K}$ whose magnitude is given by the wave number $K = 2\pi/\lambda$, where λ is the phonon wavelength (the relevant coordinate system will be discussed later).

$$\vec{p} = \hbar \vec{K} \quad (2.2)$$

$\vec{K}$ points in the direction of phonon propagation. $\vec{p}$ has the dimension of momentum but is different from its classical analog. "Longitudinal phonons" are not correlated with an inertial mass. Conservation of energy and momentum is expressed in terms of selection rules in quantum-mechanical theory. The only admissible transitions between phonon states are those which obey the rules:

$$\sum_i E_i = \text{const (1)} \quad \text{and} \quad \sum_i \vec{p}_i = \text{const (2)}$$

2.1.1 Frequency and Wave Vector

Vibrations are characterized by their wavelength λ and their period Δt. Equations of vibrations, however, are more easily handled by application of the following inverse quantities

angular frequency $\quad \omega = 2\pi/\Delta t$

wave number [1] $\quad K = 2\pi/\lambda$

Due to Eqs.(2.1) and (2.2), the quantities ω and K describe the energy and momentum respectively, of phonons. Both quantities, ω and K, are connected by the phonon velocity v. For isotropic, structureless solids it holds

$$v = \lambda/\Delta t = \omega/K = \text{const.} \tag{2.4}$$

In this case $\omega \propto K$, with no dispersion.

For systems with discrete mass points at distances d (e.g., a lattice), the Eq.(2.4) generally is not valid. When the wavelength approaches the dimension of d, v is a function of ω or K. Dispersion occurs when ω is a nonlinear function of K. The shortest meaningful wavelength is $\lambda = 2d$. In this case, standing waves establish and $v = 0$. A specific dispersion relation holds for each vibrational mode (e.g., longitudinal or transverse vibrations; see Section 2.3).

In the simple Eq.(2.4) the vectorial nature of K has not been taken into account. $\vec{K}$ is defined in a coordinate system of the reciprocal space and has the direction of the phonon velocity vector $\vec{v}$. For polymers a tetragonal system of unit cell vectors can be assumed to characterize the reciprocal lattice in the K-space. There is no direct connection between the reciprocal lattice and the lattice of vibrating masses. It is an abstract, but very useful lattice, where each point marks a solution of equations of vibrations. The K-representation is a one- or multi-dimensional display of basic vibrations of a system. It supplies an easy method of evaluating the number of possible vibrations.

Brillouin Zone. There is a certain volume in the K-space, which is necessary and sufficient to describe all vibrations of a mode in a solid. It is called the Brillouin zone, whose confinements are given by:

$K_{min} = 0 \quad$ (for $\lambda \to \infty$) and

$K_{max} = \pi/d \quad$ (for $\lambda = 2d$).

(A wavelength shorter than the double distance d between two vibrating particles is meaningless). K_{max} represents the connection to the real lattice by the lattice constant d. For anisotropic structures the components $\{K_{max}\}$ are anisotropic in the K-space. In Figure 2.1, a schematic plot is shown which demonstrates the connection of the K-lattice with the real lattice. For simplicity, the one-dimensional case of vibrating masses is used. For a linear chain the vibrations at different wavelengths λ are transposed to points of a K-coordinate within the Brillouin zone.

---

[1] The application of the inverse vibration time, the frequency, is common practice. Less popular is the inverse vibration length, the wave number K. The factor 2π is incorportated for facilitating calculations with geometrical or exponential functions, e. g., $\sin 2\pi/\lambda$.

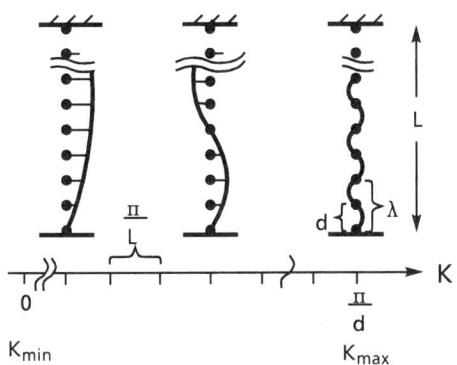

Fig. 2.1. Vibrations of a linear chain of length L with mass points at a distance d. Vibrations of different wavelengths λ are transposed to points of a K-coordinate (see Eq. 2.21).

2.1.2 Definitions of Vibrations

The first step in describing vibrations is the introduction of a linear oscillator of frequency ω. Its energy is determined by the mean number of phonons which are activated at a given temperature. An oscillator in this sense is a carrier, which can be excited to a temperature-dependent number of phonons. The next step is the extension to a three-dimensional oscillator which is able to vibrate in all directions of a body. The last step is the introduction of oscillators with different frequencies due to a frequency spectrum which is controlled by the microstructure and force constants of the solid and the mode of vibration.

This makes clear that, basically, vibrations can be described by phonons which propagate in different directions and have a material-specific frequency spectrum. The situation is more complicated for polymers since different force constants and a special structure control the type and propagation of phonons. Due to their anisotropic binding structure and their chain conformation, different vibrational modes occur in polymers. Several definitions are commonly used:

Vibrations Defined by their Force Constant. The terms stretching phonons, bending or torsional phonons symbolize the modes of vibration with respect to the force constant.

- Stretching vibrations: caused by linear force constants from covalent or Van der Waals binding forces.
- Bending vibrations: caused by valence angle deformations of the bonds.
- Torsional vibrations: caused by deformations of the azimuthal angles of the bonds.

Polarization Related to the Chain Axis. Another definition is used which distinguishes the modes of vibration by their polarization relative to the molecular chain:

- Longitudinal vibrations: parallel to the chain direction.
- Transverse vibrations: perpendicular to the chain direction.

This definition is applicable to aligned polymers. Amorphous polymers are entangled and only microdomains of such orientations exist. However, this description is a rather good approximation to thermal properties, which involve vibrations without phonon flux (e.g., specific heat, thermal expansion). Longitudinal and transverse vibrations can be described in a curved coordinate system which traces the curvature of a polymer chain. This definition of polarization is used in the subsequent considerations (The first and second definitions are closely correlated, and both can be used to describe the same vibrating mode; see Section 2.2).

Polarization Related to the Direction of Propagation. One more definition should be mentioned which relates vibrations or phonons to their direction of propagation.

- Longitudinally polarized phonons : vibrations parallel to the direction of phonon propagation.
- Transversely polarized phonons : vibrations perpendicular to phonon propagation.

This definition is not very useful for amorphous polymers since, during phonon propagation, scattering and change of directions occur (see Subsection 2.3.5).

2.1.3 Frequency Spectra

The lattice of a solid and the vibrational modes control the frequency distribution of oscillators. The number of oscillators between ω and $\omega+d\omega$ is called the density of states $D(\omega)$. The total number of linear oscillators, integrated over all possible frequencies, is equal to the number N of mass points involved. In case of three-dimensional oscillators, the total number is $3N$. This will be discussed in Section 2.4 for models applied to polymers: isolated chain, isolated stiff chain and three-dimensional arrays. Generally, the contributions from several vibrating modes or different configurations are overlapping.

2.1.4 Temperature Dependence

The energy of an oscillator is a function of temperature, and is given by its number of phonons. This quantity is given by the probability of states which is calculated by using Bose-Einstein statistics. It is a quantum mechanical version of the classical Boltzmann statistics which takes into account the quantum nature of phonons. This is especially important at low temperatures at which only a small number of phonons is excited. The probability of states gives the mean number of phonons with frequency ω which are excited at a temperature T. It correlates the energy of an oscillator to the temperature.

2.1.5 Vibrational- and Binding Energies

The anisotropic binding- and lattice structures of polymers control the phonon propagation and this is different at high and low vibrational energies. At low temperatures only low-energy phonons are activated whose wavelengths exceed mean chain distances. Thus, phonons propagate nearly equally perpendicular to the chain direction. The phonon propagates in three dimensions, although anisotropically. Low-energy vibrations are controlled mainly by the Van der Waals potentials. At high temperatures the phonon wavelengths become short, and the high-energy phonons propagate along the chains only. At a sufficient high temperature, the thermal energy of chain vibrations becomes so high that the Van der Waals binding forces are negligible. In terms of thermodynamics, it can be treated as an isolated linear chain. (The low-energy phonons are still activated, but they are no more dominant at high temperatures). The switch-over in variance of phonon propagation occurs at a temperature, at which phonons with wavelengths just equal to the mean chain distance are activated. This concept was developed by Tarasov [2.1] and extended by H. Baur taking into account the stiffness of molecular chains and introducing one- and three-dimensional arrays of stiff chains [2.2 to 2.5]. At very low temperatures, however, only long-wave phonons are activated which have little effect on the valence angles, and the influence of bending phonons disappears.

2.2 MODES OF VIBRATION

Vibrational modes are described by the following parameters:

● **Variance of phonon propagation direction.** The preceding section made clear that vibrations of polymers can be described by three-dimensional and one-dimensional phonon propagations. They are symbolized by the indices 3 or 1.

● **Polarization relative to the chain axis.** The polarization comprises longitudinal and transverse vibrations relative to the chain axis (rotations are neglected). It describes the dimensionality of vibrations of an oscillator. The indices l and t are used for the longitudinal and the (two) transverse components, respectively.

● **Force constants.** The mode of vibration furthermore depends on the phonon type which is determined by the force constant. The following force constants are used:

β: force constant of the covalent potential (β_l, and β_t);
b: bending constant (stiffness of valence angle);
γ: force constant of the Van der Waals potential (nearest neighbors of two chains);
δ: the same for the next nearest neighbors.

For an easier description, the phonon type can be attached to the notation of phonons or other thermal parameters.

Phonon type	Subscript	Binding force	Force constant
Stretching phonons	s	covalent	β (β_l, and β_t)
	sw	Van der Waals	γ, δ
Bending phonons	b	valence angle stiffness	b

Two examples are given to explain the application of the subscripts defined in this section:

● ω_{3sw}: frequency of stretching phonons (s) which propagate in three dimensions (3) and are driven by Van der Waals forces (w);
● D_{1b} : density of states of bending phonons in a linear chain (transverse bending vibrations).

Some examples of quantitative values are given in Table 2.1.

Table 2.1. Force constants [N/m].

β_1	190 [2] - 230 [3]	C $-$ C bond
γ	3.2	for a PE - crystal [2.4]
δ	0.21	
b	17	related to a deflection at an atomic distance

In Figures 2.2 and 2.3 the interplay of force constants is shown. The springs characterize the stretching force constant of the covalent potential; the dashed lines characterize the stretching forces in the Van der Waals potential; the double dashes represent the dominant bending stiffness. Collective stretching vibrations of large chain segments are indicated by arrows in Figure 2.3.

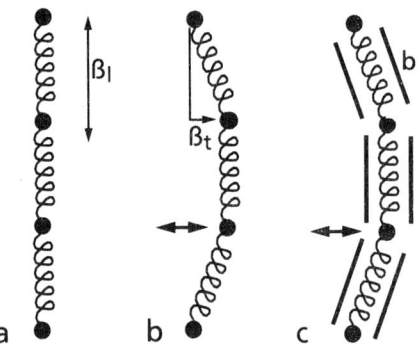

Fig. 2.2. Longitudinal and transverse vibrations of an isolated linear chain; **a)** and **b)** without bending stiffness; **c)** with bending stiffness b.

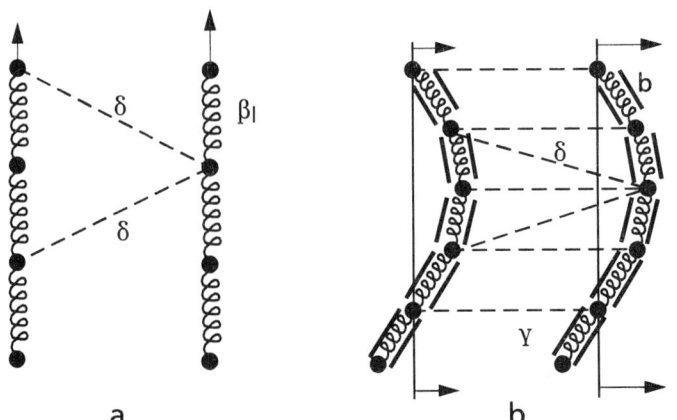

Fig. 2.3. Vibration within a lattice of aligned chains; **a)** longitudinal ; **b)** transverse.

---
[2] According to Eq. (2.11) with $E' = 120$ GPa (aligned PE).
[3] Ref. 2.19 (spectroscopic measurement).

2.3 DISPERSION RELATIONS AND PHONON VELOCITY

The correlation of ω and K is given by dispersion relations which basically relate the energy and the momentum of a phonon. The parameter linking both properties is the group velocity v of the phonons. In case of dispersion, v is a function of ω and K.

$$v = v(\omega(K)) \qquad (2.5)$$

For isotropic solids v is given by:

$$v = |\,d\omega/dK\,|. \qquad (2.6)$$

If $\omega \propto K$ no dispersion exists, and the phonon velocity is constant.

An anisotropic solid exhibits different values of ω, K and v in different directions. This is expressed as

$$\vec{v} = |\,\mathrm{grad}_K\,\omega\,| \qquad (2.7)$$

The gradient is defined by the derivative components $d\omega/dK$ in the K-space.

Dispersion relations are derived from the vibrational equations of motion. They depend on the:

- Mode of vibration.
- Variance of vibrations.
- Configuration of polymers.

Four important examples are modeled for the following considerations:
- isolated linear chain,
- three-dimensional array,
- zigzag chain,
- helical chain.

2.3.1 Isolated Linear Chain

Longitudinal Vibrations. They are described by the force constant β_l. (The zigzag structure of polymer chains and the influence of bending phonons will be considered in Section 2.3.2). For further treatment the approximation is made that due to the force constant β_l longitudinal vibrations are described by stretching phonons only. The equation of motion of a chain with masses M at distances d_i yields the well-known correlation between ω and K for longitudinal vibrations

$$\omega = (4\beta_1/M)^{1/2} \sin(Kd_i/2) \qquad (2.8\mathrm{a})$$

with β_l: force constant for longitudinal vibrations; d_i: atomic distance in a chain. For $K = \pi/d_i$ the maximum frequency $\omega_{ls\,max}$ is achieved.

$$\omega_{ls\,max} = 2(\beta_1/M)^{1/2} \qquad (2.8\mathrm{b})$$

The reduced dispersion relation reads

$$\omega/\omega_{ls\,max} = \sin(Kd_i/2) \qquad (2.8\mathrm{c})$$

In the acoustic limit $K \to 0$ (or $\lambda \gg d_i$) the chain acts as a linear continuum with

$$\omega = (\beta_l / M)^{1/2} K d_i \qquad (2.9)$$

This yields a constant phonon velocity (sound velocity).

$$v = |d\omega/dK| = (\beta_l / M)^{1/2} d_i = \omega_{ls\,max} d_i/2 = \text{const.} \qquad (2.10a)$$

An equivalent relation is well known for a thin rod having a modulus of elasticity E' and a density ρ.

$$v \approx (E'/\rho)^{1/2} \qquad (2.10b)$$

Equations (2.10a) and (2.10b) allow the force constant β_l to be evaluated from E'. In case of linear chains, E' represents the modulus of aligned polymers.

$$\beta_l = E' M (\rho d_i^2)^{-1} \; ; \quad N/m \qquad (2.11)$$

Equation (2.8a) becomes nonlinear when K gets larger due to a smaller λ. For $K \to \pi/d_i$, which is equal to $\lambda \to 2d_i$, standing waves are established whose velocity $v = 0$. The dispersion relations and the group velocities are plotted schematically in Figures 2.4a and 2.4b.

Transverse Vibrations. Transverse vibrations of a linear chain without bending stiffness are usually described by stretching phonons. The transverse component of the force constant, β_t, has the same origin as that of longitudinal stretching phonons [4] (see Fig. 2.2). For polymers the situation is more complicated. A linear polymer chain has an intrinsic bending stiffness, and bending forces (b-phonons) dominate. The approximation is made that transverse vibrations are described by bending phonons only. If the transverse vibrations are controlled only by the bending force constant b, a dispersion relation different from Eq. (2.8) holds

$$\omega = 2(b/M)^{1/2}(1 - \cos Kd_i) \qquad (2.12a)$$

For $K = \pi/d_i$

$$\omega_{lb\,max} = 4(b/M)^{1/2} \qquad (2.12b)$$

The reduced dispersion relation reads

$$\omega/\omega_{lb\,max} = \frac{1}{2}(1 - \cos Kd_i) \qquad (2.12c)$$

In the acoustic limit $K \to 0$

$$\omega \simeq d_i^2 (b/M)^{1/2} K^2 \qquad (2.13)$$

Bending phonons suffer dispersion even in the acoustic approximation. This equation corresponds to flexural waves of a stiff rod. In the acoustic limit the phonon velocity is proportional to K.

---

[4] For a linear chain without bending stiffness a similar dispersion relation holds in case of transverse vibrations with a force constant β_t. The value of β_t is obtained from an analogous relation of Eq.(2.11) by substituting the shear modulus G of aligned polymers.

$$v \approx 2 d_i^2 (b/M)^{1/2} K \qquad (2.14)$$

The dispersion relation and the phonon velocity v of a stiff chain are shown schematically in Figures 2.4c and 2.4d. The group velocity of bending phonons runs through a maximum. The velocity is zero, both for $K=0$, which corresponds to $\lambda \to \infty$, where bending forces are zero and at the boundary of the Brillouin zone, $K = \pi/d_i$ where standing waves are established. The dashed lines mark the acoustic approximation $\lambda \gg d_i$. A more detailed description of transverse vibrations takes into account the interchain forces (see Fig. 2.12).

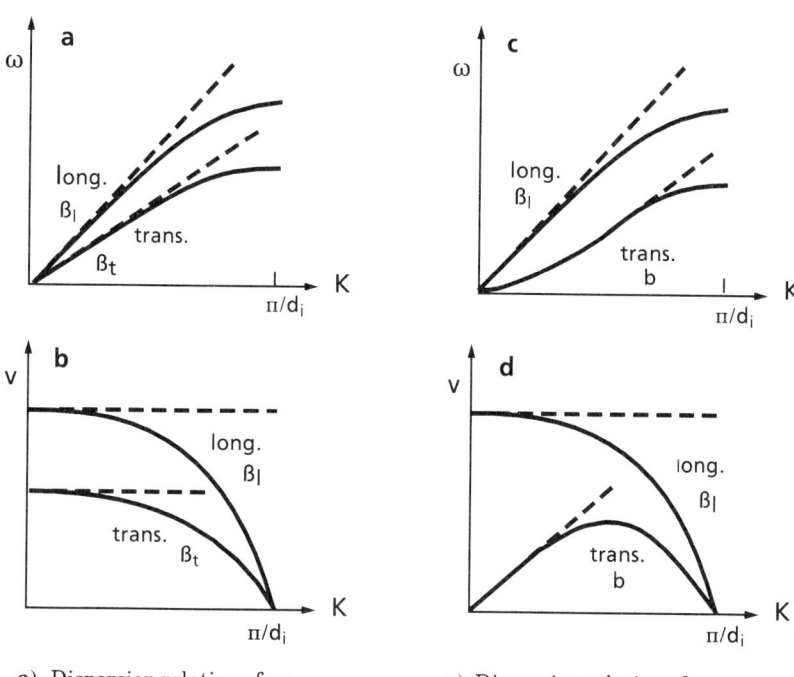

Fig. 2.4 a) Dispersion relation of an isolated, non stiff chain;
b) phonon velocity v.
c) Dispersion relation of an isolated stiff chain;
d) phonon velocity v.

2.3.2 Isolated Zigzag Chain

For linear chains, vibrations in both directions perpendicular to the chain axis are nearly equal, i.e., degenerate. This is not true for planar zigzag chains which are more representative of polymers. Different force constants act parallel and perpendicular to the zigzag plane. In addition, torsional vibrations are possibly represented by a separate dispersion branch.

A mass pair is used as a unit cell for characterizing a zigzag structure. This bisects the first Brillouin zone of a linear chain: $K \to K/2 = \pi/2d = \pi/D$. The principle of mirroring dispersion relations is plotted schematically in Figure 2.5. The dispersion relations of a zigzag chain according to Baur [2.5] are represented in Figure 2.6 (see also Kirkwood [2.6] and Pitzer [2.7]). For comparison, the dispersion relations of a linear stiff chain are added (dashed curves). They are mirrored at the bisected K-value of the first Brillouin zone, thus yielding optical branches (see Section 2.3.6). The presentation in Figure 2.6 is semischematic and illustrates the relative positions of curves.

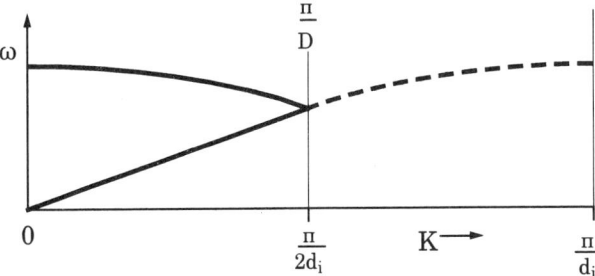

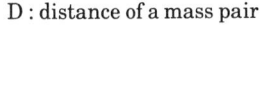

Fig. 2.5. Mirroring of dispersion relations.
d_i : atomic distance;
D : distance of a mass pair.

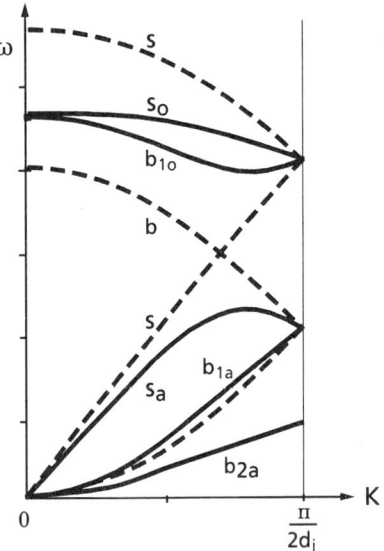

Fig. 2.6. Dashed curves : dispersion relations of a stiff linear chain;
s : longitudinal branch,
b : transverse branch.
Solid curves: dispersion relations of a planar zigzag chain;
s_a: longitudinal acoustic branch,
b_{1a}: transverse branch in the zigzag plane,
b_{2a}: transverse branch perpendicular to the zigzag plane,
s_0, b_{10}: optical branches.

2.3.3 Isolated Helical Chain

Several bulky polymers exhibit a helical structure because of steric hindrances. Additional parameters characterizing these structures are the number n of mass points per revolution and the number m of revolutions required to identically repeat the structure. The dispersion relations of helical chains are plotted in Figure 2.7 for m=1 and different values of n. In the case of a helical structure it is much clearer to substitute the wave number K of the abscissa by the phase-angle δ between neighboring vibrating mass points. Otherwise, in K-space, boundaries of the Brillouin zones would get smaller with increasing n and mirroring would lead to a mixup of curves. For example, the curves of Figure 2.7d with n=10 represent ten Brillouin zones in the K-representation. It was shown in Reference [2.8] that there exist 3n-4 optical and 4 acoustic branches for m=1.

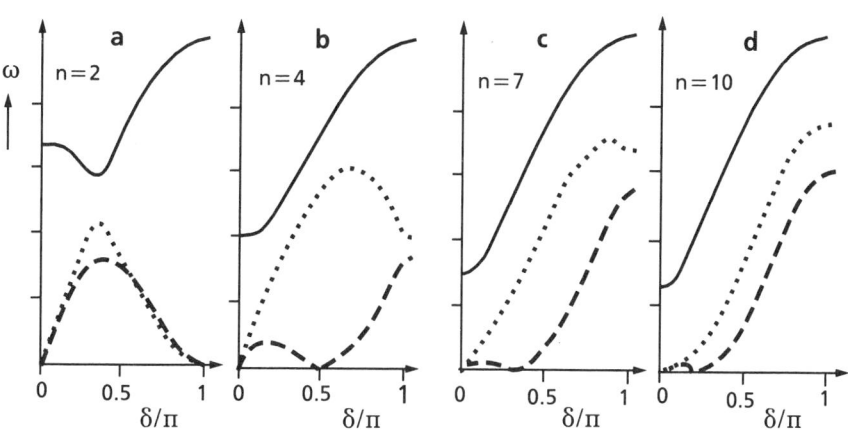

Fig. 2.7. Dispersion relations of isolated helical chains as a function of the phase-angle δ. n is the number of mass points per revolution. It is taken one revolution (m=1).
Branches: ··· longitudinal; – – transverse; ─── optical

It is evident from Figure 2.7 that optical branches start at successively lower frequencies if n is increased.[5] Helical molecules become softer if more atoms are involved in one revolution. It is unusual that optical phonons exist in the frequency range of acoustic phonons; usually they are restricted to high frequencies. In Figure 2.20 it will be shown that for helical structures the density of states $D(\omega)$ gets consequently higher with increasing n at low frequencies. The shift to lower frequencies results in a higher specific heat at low temperatures. Experimental results proved this prediction for polymers with helical structures (see Figure 3.13a for PTFE).

2.3.4 Three-dimensional Propagation

Long-wave phonons are not restricted to propagate along a chain. Spatial phonon propagation is possible by means of Van der Waals interactions between chains. They are taken into account by additional terms of the dispersion relations. In this case, dispersion relations depend on the directions of phonon propagation with respect to the polymeric chain axes. It is therefore instructive to illustrate spatial phonon propagation in the following coordinates z, ϕ, ξ which are defined in a model of aligned polymers with z along the chain axes. The plane of transverse vibrations of a chain is defined by the azimuthal angle ϕ=0. Longitudinal vibrations along a chain are given at the polar angle ξ=0 (or π).

Longitudinal Components. Besides the primary contribution of the intrachain force constant β_1, the interchain force constant δ has to be included (δ is the force constant of next nearest atoms in neighboring chains; see Fig.2.3a). Both contributions are driven by stretching phonons which obey the dispersion relation $\omega \propto K$

$$\omega = C(\beta_1, \delta) K \qquad (2.15a)$$

---

[5] The curves with n=2, m=1 correspond to zigzag chains, when mirrored at $\delta/\pi = 0.5$ (see Figure 2.6).

The term C depends on the force constants and is a geometrical function of ξ only [2.2]. The dispersion relation for various angles ξ is semischematically plotted in Figure 2.8. The largest slope (velocity) exists in chain direction ($\xi=0$). Perpendicular, the phonon velocity, driven by the Van der Waals potential, is small.

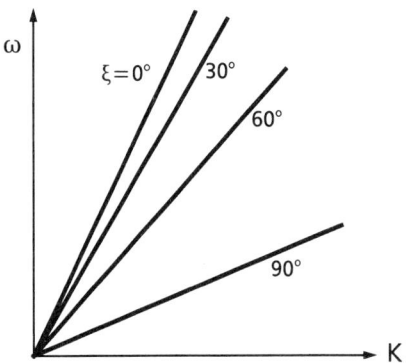

Fig. 2.8. Dispersion relations of spatial phonon propagation when chains vibrate longitudinally. (stretching phonons in the acoustic approximation) [2.2].

Transverse Components. There are two terms for transverse vibrations, the second for bending phonons (bending constant b) and the first for stretching phonons arising from Van der Waals interchain interactions (force constants γ, δ; see Fig.2.3b). For bending phonons $\omega \propto K^2$; for stretching phonons $\omega \propto K$.

$$\omega = \left[A(\gamma,\delta) K^2 + B(b) K^4 \right]^{1/2} \qquad (2.15b)$$

The terms A and B depend on the force constants and are, in addition, geometrical functions of ϕ and ξ [2.2]. The amount of A relative to B depends on the direction of phonon propagation. The maximum contribution of A occurs for $\phi = 0°$ when the phonon propagation takes place in the plane of the transverse vibrations. For propagation perpendicular to this plane ($\phi = 90°$) there is less chain interaction and the dependence on ξ is small. This is illustrated in Figures 2.9a,b in the acoustic approximation ($K \to 0$).

The following features are worth mentioning:

- **Longitudinal** phonon velocity $v = d\omega/dK$ is maximum in the chain direction
  ($\xi = 0$) (see Figure 2.8).

- **Transverse** phonon velocity, by contrast, is maximum perpendicular to the chain axis ($\xi = 90°$) (see Figures 2.9a,b);
- a large dependence on ξ exists for $\phi = 0$ (see Figure 2.9a) and
- a small dependence on ξ exists for $\phi = 90°$ (see Figure 2.9b).

However, this is true only in the acoustic approximation. At higher values of K (and ω) the phonon propagation is dominant along the chains. The contribution from the first term in Eq.(2.15b) gets smaller. (Vibrations of an isolated chain remain which are described by the Tarasov- Baur approximation).

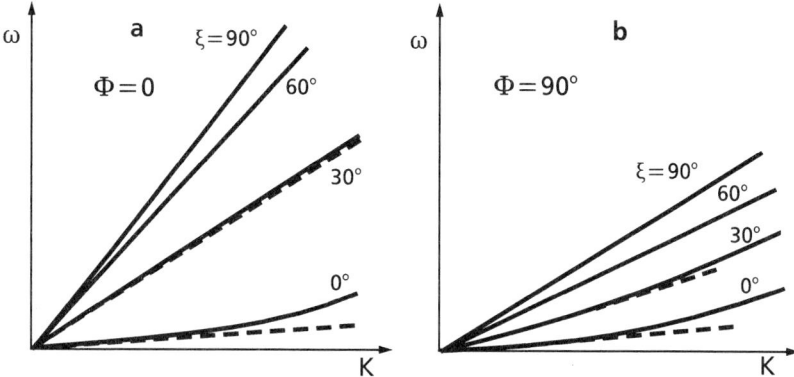

Fig. 2.9. Dispersion relation for spatial phonon propagation of transverse vibrations; **a)** in the plane of vibration ($\phi = 0$); **b)** perpendicular to the plane of vibration ($\phi = 90°$) (dashed lines: acoustic approximation).

2.3.5 Phonon Conversion

There is some probability for the conversion of vibration modes; e.g., conversion of stretching phonons to bending phonons. This is especially true for amorphous polymers. For these scattering processes energy and momentum must be conserved. In addition, the dispersion relations for the vibrating modes involved must be fulfilled. This restricts the conversion processes. For a polymer crystal (Fig.1.3a) only two types of conversion are possible if chain stiffness is neglected:

$l \rightarrow t_1 + t_2$
$l \rightarrow t_1 + l_2$

For polymeric crystals with stiff chains again two types of conversion are possible but with different energies and momenta:

$l \rightarrow t_1 + t_2$
$t \rightarrow t_1 + t_2$

In most conversions, energetic longitudinal phonons are converted into two low-energy transverse ones. This means that at high frequency longitudinal phonons have shorter lifetimes or ranges than transverse ones [2.5].

For amorphous polymers the range of phonons is even shorter; in many cases, the decay time is on the order of several periods of vibrations. Thus, the term "phonon" is no longer valid in a strict sense for amorphous solids. As too many conversions take place, one can only speak of an average number of phonons polarized longitudinally or transversely. New theories substitute short-range phonons in amorphous materials by "fractons," which characterize localized vibrations without a propagation velocity. Their contribution to internal energy plays a role below 50K. Difficulties arise since the interaction of phonons and fractons is not known [2.9 to 2.11].

2.3.6 Optical Phonons

Experimental results revealed that even at K=0 phonons can be activated by electromagnetic waves with very high frequencies up to the optical range. This is true for cases where the dipole moments, the masses or the spring constants are alternating different for successive vibrating particles. In this case, the two neighboring particles vibrate against one another with high frequencies, irrespective of the collective vibrations of the chain. In Figure 2.10 collective acoustic vibration of an inhomogeneous chain is compared with optical vibrations. In this case, the masses M_1 and M_2 are alternating. The mutual vibrations of M_1 and M_2 possess a high internal vibrational energy but their momenta cancel each other externally, and K=0. In Figure 2.10 only transverse vibrations are displayed. For longitudinal vibrations a similar situation applies.

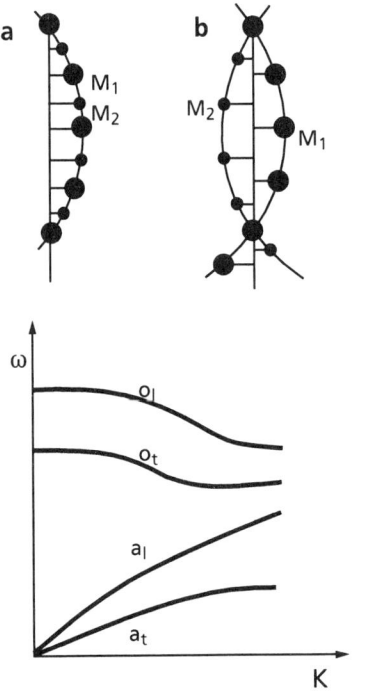

Fig. 2.10. Transverse vibrations
a) acoustic mode, b) optical mode.

Fig. 2.11. Dispersion relations within the first Brillouin zone:
a: acoustic branches,
o: optical branches,
l: longitudinal vibrations,
t: transverse vibrations.

The correlated dispersion relations are plotted in Figure 2.11. The two interesting features are that, firstly, high frequency vibrations can be activated even at K=0 and, secondly, there is usually a gap between optical and acoustic branches. The gap vanishes if the masses M_1 and M_2 (or the force constants) are equal throughout the chain. In this case Figure 2.11 represents the dispersion relations of a homogeneous chain which are mirrored in the middle of the first Brillouin zone (see Fig. 2.5). The K-values are related to the distance of a mass pair. Optical phonons are employed for interactions with electromagnetic fields. Their dispersion relations influence the refractive index and the dielectric permittivity. Their absorption frequencies reach the infrared range.

2.3.7 Experimental Measurement of Dispersion Relations

The correlation of energy (frequency) and momentum (wave vector) of acoustic and optical phonons can be detected by scattering of x-rays or neutrons. The radiation activates phonons whose energies and momenta can be analyzed. The differences of incoming and scattered energy and momentum of x-rays or neutrons are the decisive parameters. By variation of the scattering angle, a large range of ω and of the K-space can be scanned, especially with neutrons [2.11].

2.3.8 Frequency- and Temperature Dependence of Phonon Velocity

The group velocity v of phonons is given by the absolute slope of the dispersion relations. The frequency dependence is controlled by:

- the dispersion relation of a discontinuous body (e.g., a lattice);
- the phonon type (stretching or bending phonon);
- the frequency- and temperature dependence of the force constants.

Stretching phonons in a continuous, structureless body with constant force constants propagate at a velocity, which is independent of the frequency and temperature. This is true also for long-wave acoustic phonons, which do not resolve internal structures. Only short-wave phonons suffer dispersion by the discrete atomic and molecular structures.

Bending phonons, by contrast, exhibit a frequency dependence of v, even in a continuous body. As can be seen from Figure 2.4d, even for acoustic phonons $v \propto K$. The phonon velocity versus K of stretching and bending phonons in a crystalline lattice of aligned polymer chains is shown schematically in Figure 2.12. Both curves represent the velocities in chain direction.

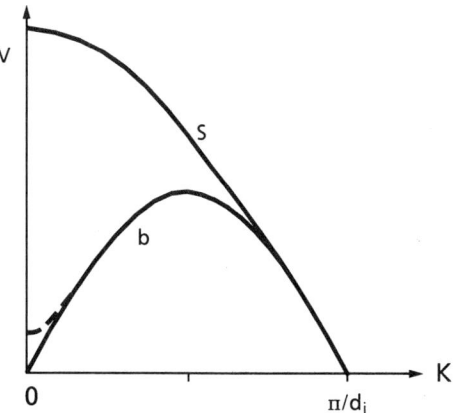

Fig. 2.12. Group velocity v versus K of phonons in a crystalline lattice of aligned chains:
  s: stretching phonons,
  b: bending phonons.

At K=0, stretching phonons have the highest velocities while for bending phonons $v=0$. The latter reflects the fact that at very long wavelengths no bending forces arise. Since transverse vibrations are influenced by Van der Waals forces of neighboring chains a small component of stretching phonons is overlapped near K=0, which causes $v > 0$ at K=0. This contribution is marked by a dashed line in Figure 2.12. At $K = \pi/d_i$ (end of the first Brillouin zone) both velocities approach zero and standing waves are established. It can be shown that bending phonons become dominant in a polymer chain above $\approx 10^{11}$ Hz [2.3]. Some information is obtained by considering v in various ranges of frequencies or wavelengths:

Acoustic Phonons ($\lambda \gg d_i$): frequency 0 to 10^{10} Hz.
This range is activated at very low temperatures. Mainly stretching phonons are activated whose dispersion is very small, with $v \approx$ const. The value of v depends on the magnitude of the force constants. The velocity of stretching phonons in a covalently bonded chain is about 10 times higher than for interchain vibrations in the Van der Waals potentials.

Thermal Phonons ($\lambda \geq d_i$): frequency 10^{12} to 10^{14} Hz.
Thermal phonons are activated at higher temperatures. The phonons suffer strong dispersions which, however, are different for stretching and bending phonons. The velocities are small, and at the boundary of the Brillouin zone only standing waves occur.

Optical Range: infrared range.
Optical branches of dispersion relations of simple crystalline lattices are rather flat in many cases (see Fig. 2.11). Therefore, the velocity of optical phonons is low. By contrast, optical phonons of helical polymers show a strong slope of dispersion relations, which results in a high velocity (see Fig. 2.7).

A temperature dependence of phonon velocity arises from the temperature dependence of the force constants. For polymers this dependence originates in viscoelastic processes which are more easily described by macroscopic elasticity moduli than by the potential theory. The force constants used in the acoustic approximation in Section 2.2 can be expressed by the elasticity and shear moduli (see Eq.(2.11)). In the acoustic approximation ($\omega \leq 10^{10}$ Hz) mainly stretching phonons are activated whose longitudinal velocity components are described by the elasticity modulus $E'(\omega,T)$ and the density ρ.

$$v_l \approx \left(\frac{E'}{\rho}\right)^{1/2} \tag{2.16}$$

This formula is valid for a rod which is thinner than the wavelength. For thicker rods the Poisson's ratio μ takes into account the deformation constraints.

$$v_l = \left(\frac{E'(1-\mu)}{\rho(1+\mu)(1-2\mu)}\right)^{1/2} \tag{2.17}$$

The transverse component of sound (or phonon) velocity is given by the shear modulus $G(\omega,T)$.

$$v_t = \left(\frac{G}{\rho}\right)^{1/2} \tag{2.18}$$

The frequency- and temperature dependence of phonon velocities is determined by that of the moduli. Equations (2.17) and (2.18) are applicable to both isotropic and oriented polymers if the isotropic and anisotropic moduli are known. The moduli are nearly constant below 30K and they, besides, do not heavily depend on the chemical composition of polymers (see: Asymptotic low-temperature behavior in Section 7.7). This is reflected in the small ranges of the phonon (or sound) velocities at 4.2K for many polymers (see Table 2.2). Deviations from the usual values occur for POM, which has a higher modulus, and for PTFE, which has a lower modulus due to the soft helical structure. Some experimental results are given in Figure 2.13.

Table 2.2. Ranges of sound velocity $v \cdot 10^3$ [m/s] at 4.2K.

Morphology	longitudinal	transverse
amorphous	2.5 to 3.5	1.4 to 1.9
semicrystalline	3.7 to 4.3	1.7 to 2.2
cross-liked	3.7 to 4	1.7 to 2.1

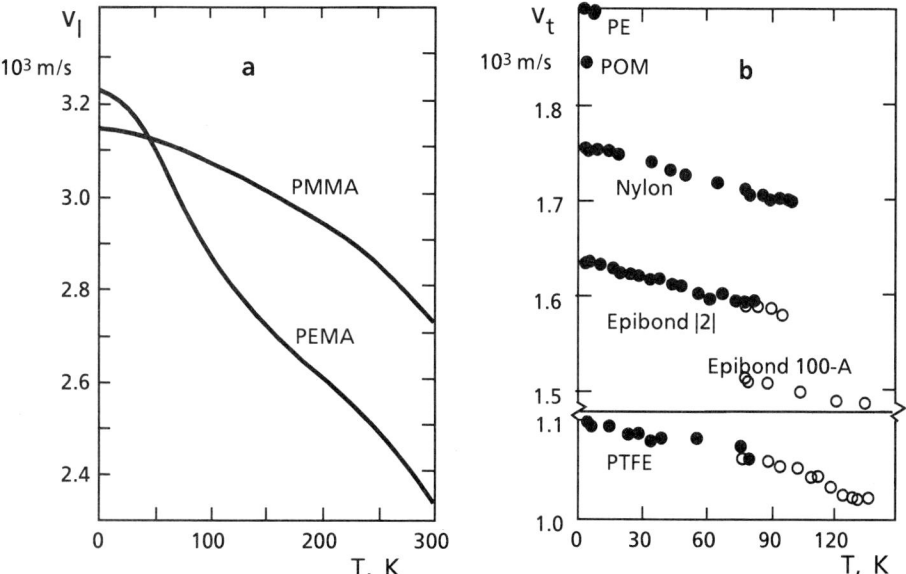

Fig. 2.13. Sound velocity versus temperature of several polymers; a) longitudinal velocity b) transverse velocity [2.13].

The sound velocities entered in Figure 2.13 or in Table 2.2 are not the pure velocities of stretching or bending phonons. They are average values of a mixture of different phonon types and vibrating modes. They represent average longitudinal and transverse components.

2.4 DENSITY OF STATES

The density of states $D(\omega)$ is defined as the number dN of possible oscillators within a frequency interval ω to $\omega + d\omega$

$$D(\omega) = \frac{\text{number of oscillators}}{\text{frequency interval}} = \frac{dN(\omega)}{d\omega} \qquad (2.19)$$

The total number N of vibrations can be normalized to the unit volume or unit length of chains or to a mole (see Section 2.4.9). The oscillating systems are assumed to consist of discrete mass points arranged in a specific structure (e.g., linear chain, lattice). For each of these arrays a specific density of states exists. Several methods are applied for the determination of D.

Direct Method. This method makes use of the fact that the projection of the dispersion curve onto the ω-axis yields $D(\omega)$. In Figure 2.14 an example is plotted (the axis of $D(\omega)$ is rotated by 90°).

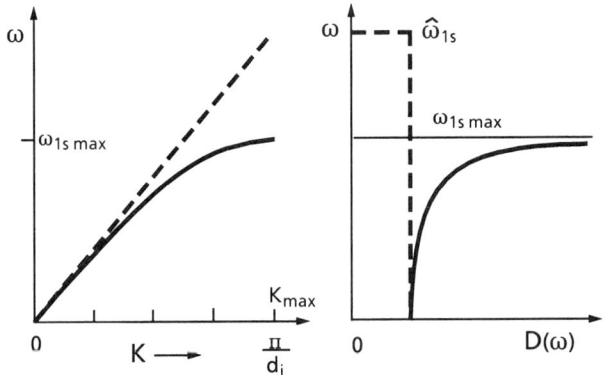

Fig. 2.14. Density of states $D(\omega)$ obtained from projecting the dispersion relation onto the ω-axis, (dashed lines: acoustic approximation).

The maximum frequency $\omega_{1s\,max}$ of a discrete oscillating system is given by the lattice constant or the Debye temperature. The frequency $\hat{\omega}_{1s}$ is an auxiliary parameter which is explained in Subsection 2.4.4.

Standing Waves Method. The first step consists in counting the number of standing waves $dN(\lambda)$ within a range λ to $\lambda + d\lambda$. The second step is to correlate λ and ω by dispersion relations. For the first step the K-vector is normally used. The number $dN(K)$ of possible vibration states is counted within a range K to $K + dK$.

$$D(\omega) = \frac{dN(K)}{dK} \frac{dK}{d\omega} = \frac{dN(K)}{dK} v^{-1} \qquad (2.20)$$

A simple example of two-dimensional vibrations is given in Figure 2.15. An area of discrete points of possible solutions of vibrational equations in a two-dimensional K-space is plotted. The number of points within the ring gives $dN(K)$. Usually only positive values of K are taken.

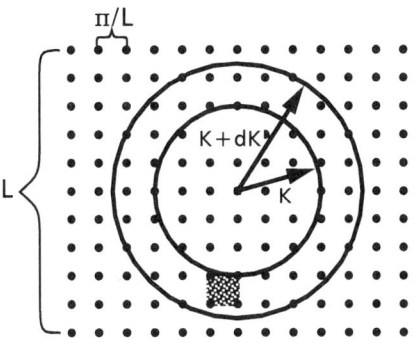

Fig. 2.15. An area with discrete mass points in the K-space is scanned within a ring of K and K+dK.

It can be easily shown that, per degree of polarization, the total number of K-values which solve the vibrational equations is equal to the number N of mass points involved. This rule can be used for calculating the distance of K-values. Two examples will be given:

Periodic Linear Chain (with N_1 mass points per unit length L at a distance d_i). A number N_1 of K-values exists which constitute solutions within the Brillouin zone from $K=0$ to $K_{max}=\pi/d_i$. The distance between K-values is therefore

$$\Delta K_1 = \frac{1}{N_1}\frac{\pi}{d_i} = \frac{\pi}{L} \tag{2.21}$$

n-dimensional Isotropic Propagation. An oscillator which propagates in n dimensions takes a K-space

$$\Delta K_n = \left(\frac{\pi}{L}\right)^n = \left(\frac{\pi}{N_n d_i}\right)^n \tag{2.22}$$

Example: each vibration in an isotropic three-dimensional body of unit volume V spans a discrete K-volume

$$\Delta K_3 = \left(\frac{\pi}{L}\right)^3 = \pi^3/V$$

The K-dependent number N(K) of vibrations can be calculated from the total, continuous K-space, K_{tot}, spanned by oscillator systems, which is divided by the discrete space ΔK_n of one oscillator.

$$N_n(K) = K_{tot}/\Delta K_n \quad \text{and} \quad \frac{dN_n(K)}{dK} = \frac{dK_{tot}}{dK}/\Delta K_n \tag{2.23}$$

Two examples will be added for illustrating Eq.(2.23):

Linear Chain: for a one-dimensional K-line $K_{tot}=K$ and $\Delta K_1=\pi/L$

$$N_1(K) = \frac{L}{\pi}K \quad \text{and} \quad \frac{dN_1(K)}{dK} = \frac{L}{\pi} \tag{2.24a}$$

Isotropic Solid: three-dimensional K-sphere

$$K_{tot} = \frac{4}{3}\pi K^3 \quad ; \quad \Delta K_3 = \left(\frac{\pi}{L}\right)^3 = \pi^3/V$$

Assuming only positive K-vectors

$$K_{tot} = \frac{\pi}{6} K^3$$

$$N_3(K) = \frac{V}{6\pi^2} K^3 \quad \text{and} \quad \frac{dN_3(K)}{dK} = \frac{V}{2\pi^2} K^2 \tag{2.24b}$$

2.4.1 Linear Chain

Longitudinal Vibrations. A linear chain of unit length L with equal masses is assumed ($L \gg \lambda$). Interactions with neighboring chains are neglected. From Eqs.(2.20) and (2.24a) one gets

$$D_{1s}(\omega) = \frac{dN_1}{dK} v^{-1} = \frac{L}{\pi v} \tag{2.25a}$$

Assuming only stretching vibrations and a constant velocity v it holds:

$$D_{1s}/L \approx \text{const.} \tag{2.25b}$$

Therefore, the density of states per unit chain length is constant for stretching phonons. An analogous result is obtained if $D_{1s}(\omega)$ is related to the number N_{1s} of mass points of a chain.

$$D_{1s} \approx \frac{N_{1s} d_i}{\pi v} \approx 2 N_{1s}/(\pi \hat{\omega}_{1s}). \tag{2.25c}$$

The last term of Eq.(2.25c) results from Eq.(2.10a), where $\hat{\omega}_{1s}$ is an auxiliary maximum frequency shown in Figure 2.16 and d_i is the atomic distance. A different function results if v is calculated exactly from Eq.(2.8c).

$$D_{1s} = \frac{2N_{1s}}{\pi} (\omega_{1s\,max}^2 - \omega^2)^{-1/2} \tag{2.26}$$

Both functions are plotted in Figure 2.16a. Equation (2.26) yields a singularity at $\omega_{1s\,max}$. The singularity is avoided in the acoustic approximation (dashed lines). The area below both curves are made equal by normalizing to N_{1s}. The maximal frequency $\hat{\omega}_{1s}$ results from normalization and has no physical meaning (see Subsection 2.4.4).

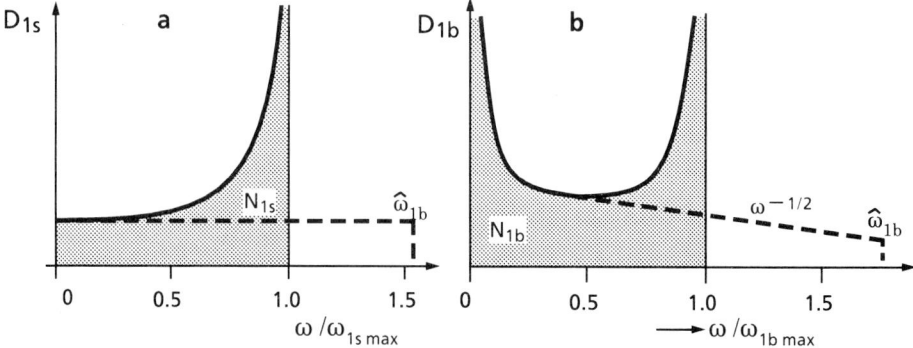

Fig. 2.16. Density of states of a linear chain: **a)** longitudinal vibrations, **b)** transverse vibrations. ▬ exact solution, ▬ ▬ acoustic approximation.

PHONONS

Transverse Vibrations. The term $dN(K)/dK$ is the same as for longitudinal vibrations; however, the dispersion relation and the phonon velocity are different for transverse vibrations. From Eqs.(2.25a) and (2.14a) and in the acoustic approximation it results

$$D_{1b}(\omega) = \frac{N_{1b}}{\pi} (\widehat{\omega}_{1b} \omega)^{-1/2} \tag{2.27}$$

The exact calculation is omitted but is plotted in Figure 2.16b [2.5].

2.4.2 Three-dimensional Isotropic Solid

Vibrations are described by stretching phonons. Due to a spherical shell in the K-space it follows from Eqs.(2.20) and (2.24b)

$$D_{3s}(\omega)/V = \frac{K^2}{2\pi^2} \frac{dK}{d\omega} = \frac{\omega^2}{2\pi^2 v^3} \tag{2.28a}$$

A schematic diagram is shown in Figure 2.17. $\omega_{3s\,max}$ is the maximum frequency. A complication arises from polarization, since the velocities of longitudinal and transverse phonons are different. One can define a density of states separately for longitudinal and transverse phonon propagations.

$$D_{3s}(\omega)/V = \frac{1}{3} \frac{\omega^2}{2\pi^2 v_l^3} + \frac{2}{3} \frac{\omega^2}{2\pi^2 v_t^3} \equiv \frac{\omega^2}{2\pi^2 \overline{v^3}} \tag{2.28b}$$

Normalization is done for one longitudinal and two transverse branches. Usually, the transverse velocity v_t and the longitudinal velocity v_l are averaged. Since $v_l > v_t$, the approximation holds: $1/\overline{v^3} \approx 2/3v_t^3$

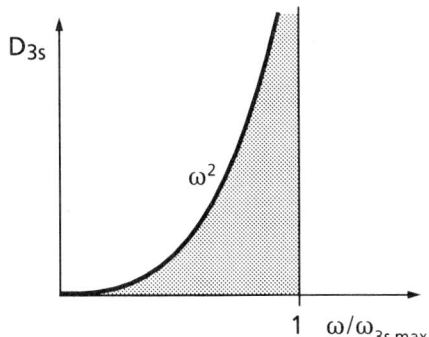

Fig. 2.17. Density of states of an isotropic solid.

2.4.3 Three-dimensional Anisotropic Polymers

For a very anisotropic lattice of aligned polymer chains the calculation of $D(\omega)$ is much more complicated. Several types of phonons are involved:

- Stretching phonons in covalently bonded chains.
- Stretching phonons between chain segments vibrating in the Van der Waals potential.
- Bending phonons due to intrinsic chain stiffness.

An approximative calculation was given by Tarasov [2.1]. H. Baur refined this theory by introducing bending phonons, which better describe transverse vibrations in polymers [2.2]. The density of states of transverse vibrations propagating in three dimensions is given by

$$D_{3b}(\omega) \propto \omega^{+1/2} \qquad (2.29)$$

(The case of one-dimensional propagation was already given by Eq.(2.27).) The following treatment subdivides longitudinal components (stretching phonons) and transverse components (bending phonons) for three- and one-dimensional propagations.

Longitudinal Components. At low frequencies (or temperatures) long-wave phonons are activated which propagate in three dimensions and obey the relation: $D_{3s}(\omega) \propto \omega^2$. At elevated temperatures, above a certain frequency ω_{3s}, the short-wave phonons propagate along the chains and obey the relation: $D_{1s}(\omega) = \text{const}$. A schematic plot is shown in Figure 2.18a.

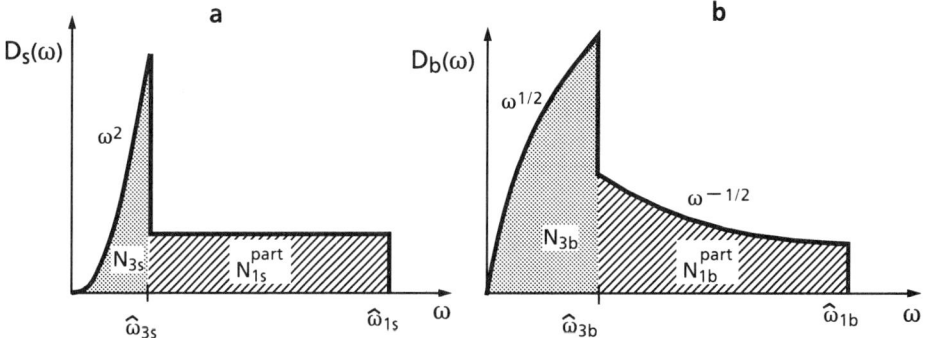

Fig 2.18. Densities of states in an aligned polymer. a) longitudinal components, b) transverse components

Transverse Components. In a stiff polymer lattice bending phonons control mainly the transverse vibrations. At low temperatures long-wave bending phonons propagate in three dimensions and obey the relation: $D_{3b}(\omega) \propto \omega^{1/2}$. At elevated temperatures short-wave phonons propagate along the bending stiff chains and obey the relation: $D_{1b}(\omega) \propto \omega^{-1/2}$. Both relations are plotted schematically in Figure 2.18b. The limiting frequencies $\hat{\omega}_{1s}$ and $\hat{\omega}_{1b}$ are auxiliary parameters, which are described in Section 2.4.4. The combination of both, the longitudinal and transverse components, describe roughly the density distribution of vibrational states (oscillators) in polymers.[6] For POM this rough calculation (acoustic approximation) is plotted in Figure 2.19 by dashed lines. It is compared with a more exact calculation by potential theory (force constants) plotted as solid lines [2.14]. It is worth mentioning that already the acoustic approximation yields a rather good result.

---

[6] The vibrations along the chains propagate in one dimension, but the oscillators vibrate three-dimensionally (one longitudinal component and two transverse components).

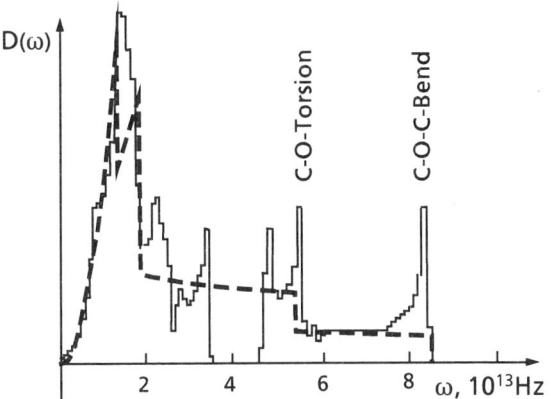

Fig. 2.19. Calculations of the density of states for POM [2.14]:
— by potential theory,
– – by acoustic approximation,
(combination of Figs. 2.18a,b).

Finally, an example is given of the density of states of a spiral (helix) polymer. Referring to Figure 2.7, four mass points (n=4) are assumed per revolution (m=1). In Figure 2.20 a plot of $D(\omega)$ is presented which shows the relative contributions of longitudinal, bending (transverse) and optical phonons [2.8]. As already expected from dispersion relations, a large contribution is made by optical phonons in the low-frequency region. This feature is specific of polymers with helical structures.

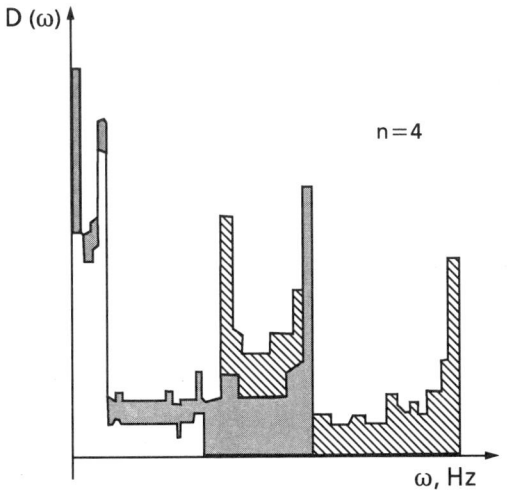

Fig. 2.20. Density of states of polymers with helical structure:

▨ longitudinal,
▧ transverse,
☐ optical.

2.4.4 Normalization

The density of states integrated from $\omega = 0$ to ω_{max} is equal to the total number N of vibrating masses per degree of freedom n

$$\int_0^{\omega_{max}} D(\omega) \, d\omega = N \tag{2.30}$$

This equation is fulfilled identically if the exact function $D(\omega)$ and the maximum frequency ω_{max} of an oscillating system are known. Usually, approximations of $D(\omega)$ are applied, and auxiliary maximum frequencies $\hat{\omega}_{nm}$ are introduced for a correct normalization to N (see Fig. 2.16). Each mass point in a body is assumed to have generally three degrees of freedom for vibrations (rotations are neglected)

$$N_3 = 3N. \tag{2.31}$$

The number N_3 of oscillations consists of N_s longitudinal and N_b transverse components

$$N_3 = N_s + N_b \tag{2.32a}$$

Due to the chain structure, N_b is twice that of N_s.

$$N_s = \frac{1}{3}N_3 \quad \text{and} \quad N_b = \frac{2}{3}N_3. \tag{2.32b}$$

In order to avoid double counting, especially in the three- and one-dimensional presentations of the Tarasov theory, the following relative normalizations are introduced:

Longitudinal Components:

$$N_s = N_{3s} + N_{1s}^{part} \tag{2.33a}$$

Transverse Components:

$$N_b = N_{3b} + N_{1b}^{part} \tag{2.33b}$$

The areas in Figures 2.18a and 2.18b are normalized in these numbers N_{nm}. The normalization can be done by incorporating a factor to $D(\omega)$ or by adjusting the auxiliary frequency limits $\hat{\omega}_{nm}$. In Figure 2.18a, e.g., the normalization holds

$$N_s = \int_0^{\hat{\omega}_{3s}} D_{3s}(\omega)\,d\omega + \int_{\hat{\omega}_{3s}}^{\hat{\omega}_{1s}} D_{1s}(\omega)\,d\omega \tag{2.34}$$

where the first term is equal to N_{3s} and the second one represents part of N_{1s}. The low frequency vibrations up to ω_{3s} of the linear chain are already included in the term N_{3s} (and similar for the transverse components). The densities of states are normalized to parameters, which allow a relative normalization to be made.

For an easier understanding, the density of states of a linear chain has been normalized to the chain length L or to their number N_{1s} of mass points per chain (see Eq.(2.26)). For real calculations the total number of vibrations in the whole body is required. Thus, a linear chain is not normalized to the number N_{1s} per chain, but to the total number, which includes all chains of a body. In a common manner the number N is related by the Avogadro's number to a mole and by the mole mass to the density.

Debye Frequencies and Debye Temperatures. The Debye frequencies $\hat{\omega}_{nm}$ can be transposed to the Debye temperatures Θ_{nm}

$$\Theta_{nm} = \hbar\hat{\omega}_{nm}/k_B \tag{2.35}$$

where k_B is the Boltzmann constant. Θ describes the temperature, at which the maximum frequency is activated. For vibrations along a covalently bonded polymer chain, the Debye temperature is on the order of $\Theta = 400K$ to $800K$. Interchain vibrations within a Van der Waals potential yield $\Theta \approx 50K$ to $150K$. At least at 150K all vibrations, driven by the Van der Waals potential, are saturated. Each vibrational mode of a polymer has its own Debye temperature. It is difficult to compile such a set of Θ_{nm}. Some information can be obtained from analyzing the specific heat. More information is available from measurements of the dispersion relations resulting from neutron scattering. Some values of $\hat{\omega}_{nm}$ and Θ_{nm} are compiled in Table 2.3 for POM [2.15]. For the calculation of specific heat (Eqs.(3.8a,b)) the values of Θ_{nm} are used for the relative normalizations.

Table 2.3. $\hat{\omega}_{nm}$ and Θ_{nm} of POM.

Index : nm	1s	1b	3s	3b
$\hat{\omega}_{nm}$; (Hz)	5.43×10^{13}	8.25×10^{13}	1.9×10^{13}	1.4×10^{13}
Θ_{nm}; (K)	415	651	145	108

2.4.5 Tunneling Systems

Tunneling vibrations dominate in amorphous polymers at very low temperatures. The energy levels of tunneling systems have a broad and rather constant density distribution [2.16]. This is assumed also for the frequency distribution of tunneling vibrations. Therefore it holds

$$D(\omega) = \text{const.} \quad (2.36)$$

This assumption is consistent with specific heat contributions from tunneling vibrations (see Subsection 3.4.2).

2.4.6 Variance and D(ω)

The frequency dependence of $D(\omega)$ for stretching phonons is correlated to the spatial variance n of an oscillating system. In the acoustic approximation

$$D(\omega) \propto \omega^{n-1} \quad (2.37)$$

one-dimensional propagation $n = 1$: $D(\omega) = \text{const.}$
two-dimensional propagation $n = 2$: $D(\omega) \propto \omega$
three-dimensional propagation $n = 3$: $D(\omega) \propto \omega^2$

Equation (2.37) also holds for the fractal dimension n when conformations of polymer chains are measured by fractons. The fractal theory of polymers yields: $n_f = 5/3$ [2.21]. (Results of specific heat over a large temperature range are consistent with the assumption of a fracton density of states).

2.5 PROBABILITY OF STATES

After the density of states (or oscillators) D(ω) has been calculated, the next question concerns the energy of an oscillator. Its calculation is given by the probability of states, which relates the mean number of phonons to the temperature. At absolute zero, only zero-point vibrations exist with a quantum-mechanical zero point energy $E_0 = 1/2\hbar\omega$. At higher temperatures more phonons are generated, the number of which is given by the parameter m. An oscillator of the frequency ω has quantized energy levels

$$E_m = (1/2 + m)\hbar\omega; \quad m = 0, 1, 2... \quad (2.38)$$

The values of E_m are analogous to the classical amplitudes of vibrations (more correctly to the squared amplitudes).

Phonons are Bose particles and obey Bose-Einstein statistics at equilibrium. The mean number of phonons with a frequency ω which are created at a temperature T is called the probability of states and is given by Bose-Einstein statistics

$$<m(\omega,T)> = [\exp(\hbar\omega/k_B T) - 1]^{-1} \quad (2.39)$$

Equation (2.39) is plotted in Figure 2.21. The mean energy of a linear oscillator with the frequency ω at a temperature T is given by

$$E(\omega,T) = <m(\omega,T)>\hbar\omega + E_0 \quad (2.40)$$

This energy is due to vibrations in one direction. If the same oscillations occur in three independent directions, a threefold amount of energy results. (The variance and polarization of vibrations usually are included in the densities of state).

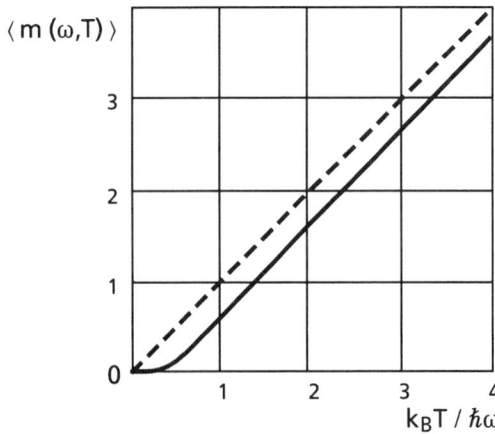

Fig. 2.21. Probability of states.
-- classical limit
—— Bose-Einstein statistics

At high temperatures, $\hbar\omega \ll k_B T$ and the classical limit $<m> = k_B T/\hbar\omega$ is approached. Thus, from Eq.(2.40) the mean energy of an oscillator becomes independent of frequency and is given by

$$E(T) = k_B T \quad (2.41)$$

If the oscillator energy is high enough, the quantized energy $\hbar\omega$ of a phonon is relatively small and causes nearly no energy step.

The probability of states generally can be used to relate the frequency to the temperature. Assuming a certain number $<m>$ of phonons, one can ask for the maximum frequency which is activated at a given temperature T. As an example, the creation of one phonon at different frequencies and temperatures is considered. The value $<m(\omega,T)> = 1$ is obtained by the following combinations of temperature and frequency:

$$\omega = 10^{11} \text{ Hz} \quad \text{at about} \quad 1\text{K}$$
$$\omega = 10^{12} \text{ Hz} \quad \text{at about} \quad 10\text{K}$$
$$\omega = 10^{13} \text{ Hz} \quad \text{at about} \quad 100\text{K}.$$

2.6 SUMMARY

(1) Vibrations are described by phonons, which comprise all vibrational parameters. They are of quantum-mechanical nature. Their energy $\hbar\omega$ and their momentum $\hbar K$ can only be changed stepwise. This feature is important to describing low-temperature thermal properties.

(2) Vibrations usually are described by the inverse vibration time (frequency) and the inverse wavelength (K-value). The K-space represents a display of possible solutions of the vibrational equations. Its connection to the real space is given by the lattice constants (e.g., atomic or chain distance).

(3) Vibrations in polymers are driven by different force constants:
    - covalent stretching forces within a chain,
    - Van der Waals stretching forces between chains,
    - bending forces by stiffness of valence angles of the bonds.

(4) The mode of a vibration is given by the force constant and the variance of phonon propagation. The mode of longitudinal vibrations is approximately described by stretching phonons; the mode of transverse vibrations by bending phonons.

(5) Vibrations propagate three-dimensionally at very low temperatures and one-dimensionally along the chains at elevated temperatures. This is described by the Tarasov theory and a theory extended by Baur.

(6) There is a distinction between the variance of phonon propagation and the variance of oscillations. Vibrations, e.g., which propagate linearly along the chains vibrate in three dimensions: one longitudinal component and two transverse components.

(7) The frequency and K-number dependences of a mode are described by dispersion relations, whose slopes yield the phonon velocity.

(8) In addition to acoustic branches, optical branches exist for oscillating systems with alternating masses or force constants. Optical, high-frequency vibrations result from vibrations of neighboring masses against one another. They possess a high internal energy but no collective momentum.

(9) Tunneling vibrations are driven by phonon absorption; small molecular units vibrate between two energy levels. They are dominant below 1K.

(10) The density of states gives the number of possible vibration states (oscillators) within ω and $\omega + d\omega$. The density of states depends on the maximum frequencies ω_{max} of a mode. Experimentally, ω_{max} can be determined from the Debye temperature Θ or from neutron scattering.

(11) The dispersion relations, the phonon velocity and the density of states depend on the force constants and variance of phonon propagation. Complications arise from the complex structure of polymers (e.g., zigzag or helix structure). In most cases, the approximation of aligned linearized chains is sufficient for calculating thermal properties.

(12) At very low temperatures, only long wavelength (acoustic) phonons are activated which do not resolve the discrete structure of polymers. Acoustic approximations describe the vibrational properties rather well. At elevated temperatures, the discrete structure has to be taken into acount.

(13) The mean number of phonons with the frequency ω activated at a temperature T is given by Bose-Einstein statistics, which is a quantum-mechanical version of the Boltzmann statistics.

2.7 REFERENCES

2.1 Tarasov,V.V.; Z. Fiz. Khim. 24, (1950) 111.
2.2 Baur, H., Kolloid Z. u.Z. Polym. 241 (1970), 1057.
2.3 Baur, H., Zeitschrift f. Naturf. 26 A (1971), 979.
2.4 Baur, H., Kolloid Z. u.Z. Polym. 244 (1971), 293.
2.5 Baur, H., Kolloid Z. u.Z. Polym. 250 (1972), 1015.
2.6 Kirkwood, J.G.; J. Chem. Phys. 7, (1939), 506.
2.7 Pitzer, K.S.; J. Chem. Phys. 8, (1940), 711.
2.8 Ya. V. Telezhenko and B.Y. Sukharevskii; J., Low Temp. Phys. 8 (2), (1982); p. 93.
2.9 Rosenberg, H.M.: Inelastic Neutron Scattering in Epoxy Resins; Phys. Rev. Let. 54; p. 704; (1985)
2.10 Tua, P.F., Puttermann S.J. and Orbach, R.; Phys. Lett. 98 A (1983), 357.
2.11 Allen, J.P.; J. Chem. Phys. 84 (8) (1986); p. 4680.
2.12 White, J.W.; in " Structural Studies of Molecules by Spectroscopic Methods," Ed. Irin; Wiley, N.Y. (1976).
2.13 Filipezynski,L.; Powlonski,Z. and Wehr,J.; in " Ultrasonic Methods of Testing Materials," Butterworth (1966).
2.14 Kitagawa,T. and Miyazawa,T.; Rept. Prog. Polym. Phys. Japan 9, (1966); p. 175.
2.15 Engeln,J.; doctoral thesis 1983, Berlin; D 83; and in "Nonmetallic Materials and Composites at Low Temperatures," Vol. 2, Plenum Press, N.Y. Eds. G. Hartwig, D. Evans, pp. 1 - 16, (1982).
2.16 Hunklinger, S. and Schickfus; in "Amorphous Solids," Topics in Current Physics, Ed. W.A. Philips, Vol. 24; Springer Press, Berlin, p. 81, (1982).
2.17 Mandelbrot B.B.: The fractal geometry of nature; W.H. Freemen Co., N.Y. (1983), p. 329.
2.18 Tasumi, M.; F. Shimanouchi and T. Miyazawa; J. Mol. Spectr. 9; p. 261; (1962).
2.19 Borges da Costa, J.A. ; Z. f. Naturf. 38a (1983); p. 1284.

General Reading

1. Reissland J.A. : Physics of Phonons ; John Wiley and Sons, New York (1973).
2. Kittel, C. ; Introduction to Solid State Physics; John Wiley and Sons, New York (1973).
3. Ziman, J. A.; Electrons and phonons; Oxfort Press; (1960).
4. White, J.W.; in "Dynamics of Solid and Liquids by Neutron Scattering," Ed. Lovesey; Springer Press, Berlin (1977).

SPECIFIC HEAT

Contents

3.	Specific heat	49
3.1	Definition and survey	49
3.2	Theory	50
3.3	Specific heat of polymers	53
3.4	Experimental results on temperature dependence	55
3.5	Dependence on crystalline content	58
3.6	General results and discussion	62
3.7	Molar heat conversion	63
3.8	Summary	64
3.9	References	64

Appendix

3A   Data

3. SPECIFIC HEAT

3.1 DEFINITION AND SURVEY

Heating a body with a mass m by an energy ΔQ leads to a material specific temperature rise ΔT.

$$\Delta Q = c\, m\, \Delta T \quad ; \quad J. \tag{3.1}$$

The coefficient c relating ΔQ to ΔT is called the specific heat. In terms of thermodynamics, the specific heat is defined by the temperature dependence of the internal energy or the enthalpy [3.1].

At constant volume, the specific heat is defined as the change of internal energy U per temperature variation

$$c_v = \partial U/\partial T_v \quad ; \quad J/(gK). \tag{3.2a}$$

At constant pressure, the specific heat is defined as the change of enthalpy H

$$c_p = \partial H/\partial T_p \quad ; \quad J/(gK). \tag{3.2b}$$

In this case a temperature variation causes, in addition, a thermal volume expansion β. The interrelation of c_v and c_p is expressed as

$$c_p - c_v \simeq T\beta^2/(\rho\kappa) \tag{3.3}$$

where ρ denotes the density and κ the compressibility. At low temperatures, β is very small for condensed phases so that

$$c_v \approx c_p = c$$

In the definitions above c, U, and H are assumed to be related to mass. Other definitions relate these properties to the volume or the molar weight.

A schematic survey of the specific heat of polymers as a function of temperature is given in Figure 3.1. The tendency can be seen that the specific heat decreases drastically at low temperatures. This is explained by a quantum-mechanical treatment. The thermal vibrations which are the carriers of heat, can be described by phonons. The quantization of the phonon energy causes a temperature dependence of the internal energy quite different from that in the classical treatment. This change is especially strong at low temperatures at which the phonon energy is small, and only few phonons are activated.

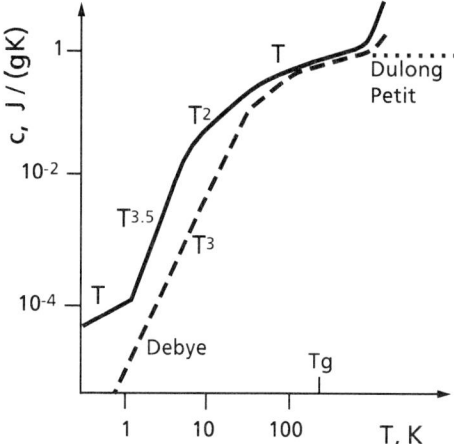

Fig. 3.1. Schematic representation of the specific heat versus temperature.
— amorphous polymers,
-- crystalline polymers.

Highly crystalline polymers obey the Debye law $c \propto T^3$ at low temperatures. At elevated temperatures roughly $c \propto T$, which is specific for linear chains. At high temperatures the Dulong-Petit law $c = \text{const.}$ is expected, but in most cases contributions from main glass transitions at T_g are superimposed.

Amorphous polymers show a different behavior:

(a) Their specific heat is remarkably higher than that of highly crystalline ones below 60K. This is due to the following reasons:

- looser packing (vibrations can be activated at relatively lower temperatures);
- contributions from tunneling vibrations below 1K;
- smeared out contributions from a background of weak glass transitions.

(b) The temperature dependence of the specific heat is different for amorphous polymers due to additional vibrational modes:

- tunneling vibrations are dominant below 1K and obey the relation $c \propto T$;
- localized mode vibrations lead to a stronger temperature dependence in the Debye region: $c \propto T^3$ to $T^{3.5}$

It is a general feature that the specific heat below 80K of most amorphous polymers is rather similar and does not depend very much on their chemical composition. This is, to a lesser degree, also true for semicrystalline polymers, when they are related to the same crystalline fraction. Some influence has been found on the conformation of the crystallites (e.g., lamella or helix).

3.2 THEORY

3.2.1 Specific Heat from Single-mode Vibrations

The basic theory is given for crystalline materials; modifications are made for amorphous polymers. The internal energy of a vibrating system is given by the product of the phonon energy $\hbar\omega$, the densitiy of states $D(\omega)$, and the probability of state $\langle m(\omega,T) \rangle$, integrated up to a maximum frequency ω_{max}

$$U(T) = \int_0^{\omega_{max}} \hbar\omega \langle m(\omega,T) \rangle D(\omega) d\omega \quad ; \quad J/g \quad (3.4)$$

SPECIFIC HEAT

$\langle m(\omega,T)\rangle$ is given by Bose-Einstein statistics and indicates the mean number of phonons with the frequency ω, excited at a temperature T (see Eq.(2.39)). For homogeneous isotropic solids one applies the densities of states for longitudinal and transverse phonon propagations (Eq.(2.28b)).

$$D(\omega) = \frac{1}{3}\frac{V}{2\pi^2 v_l^3}\omega^2 + \frac{2}{3}\frac{V}{2\pi^2 v_t^3}\omega^2 \equiv \frac{V}{2\pi^2 \bar{v}^3}\omega^2$$

where $\bar{v}^3$ is a mean value of phonon velocity, which is determined by longitudinal and transversal components (see Subsection 2.4.2).

The Debye representation of the specific heat is given by Eqs.(3.2a) and (3.4)

$$c = \partial U/\partial T = \frac{\partial}{\partial T}\frac{V\hbar}{2\pi^2 \bar{v}^3}\int_0^{\omega_{max}}\frac{\omega^3}{e^{\hbar\omega/k_B T}-1}d\omega \quad ; \quad J/(gK) \qquad (3.5a)$$

For T→0, Eq.(3.5a) yields the well-known low-temperature approximation [3.1] (Debye law)

$$c = \text{const.}\frac{1}{\bar{v}^3}T^3 \quad ; \quad J/(gK) \qquad (3.5b)$$

The constant contains normalization parameters and the maximum frequency ω_{max}. The considerable decrease of specific heat, observed at very low temperatures, is due to the quantization of the phonon energy. There is no steady change with the temperature. At absolute zero, only the basic energy $\hbar\omega_0/2$ is excited and this does not contribute to the specific heat. ω_0 is the basic frequency of the oscillator. Excitation of a next higher energy state occurs after a certain increase in temperature. Figure 3.2 shows the contributions to the specific heat from different frequencies ω of a harmonic oscillator versus temperature, which is normalized by the Debey temperature in order to obtain a material invariant representation. The ordinate is normalized to the Boltzmann constant k_B, which represents the maximum contribution to the specific heat at high temperatures. It is obvious from Figure 3.2 that the contribution from an oscillator becomes very small at low temperatures. Higher basic frequencies can only be excited at higher temperatures. This is the consequence of quantum-mechanical treatment. In the classical theory each oscillator contributes a constant value k_B to the specific heat per degree of freedom. This is the limiting value at high temperatures $T \gg \Theta$.

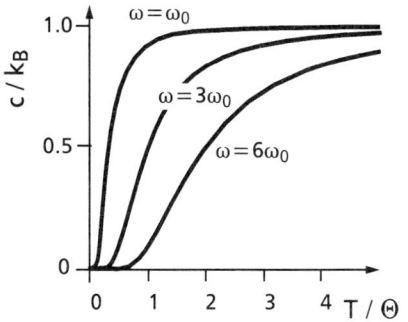

Fig. 3.2. Contributions to the specific heat by oscillators with frequencies $\omega_0, 3\omega_0,$ and $6\omega_0$, versus temperature.

3.2.2 Specific Heat from Multi-mode Vibrations

Different modes of vibrations contribute to the specific heat of polymers. Vibrations of a three-dimensional continuum are activated at low temperatures and at higher temperatures vibrations of the linear chains dominate (Tarasov theory). An improved description takes into account the bending stiffness of the chains according to Baur [3.4] (Contributions of side groups or side substituents are not considered separately in this treatment of chains). The mode of vibrations is determined mainly by the density of states $D_{nm}(\omega)$ and the maximum frequency $\hat{\omega}_{nm}$ (or the Debye temperature Θ_{nm}). Refering to Section 2.2 the subscript n denotes the variance of phonon propagation and m the type of vibration (s for stretching and b for bending). For tunneling vibrations one subscript $D_t(\omega)$ is sufficient (see Table 3.1). As shown in Section 2.7, the density of states of any mode is given by a specific power law: $D(\omega) \propto \omega^n$. This is true for any exponent n, even for noninteger ones (e.g., for bending vibrations of stiff linear chains $D_{1b} \propto \omega^{-1/2}$). The exponent n is not correlated to the subscript n.

From Eqs.(3.2a) and (3.4) one can derive a general relation for any mode with $D(\omega) \propto \omega^n$. Taking the abbreviation $x = \hbar\omega/k_B T$, a generalized Debye function can be written as [3.2]

$$g_{nm}(T,\Theta) = (n+1)(T/\theta_{nm})^{n+1} \int_0^{x_{max}} x^{n+2} e^x (e^x - 1)^{-2} dx \tag{3.6}$$

The maximum frequency ω_{max} is commonly expressed by the Debye temperature $\Theta = \hbar\omega_{max}/k_B$. Consequently, it holds that $x_{max} = \Theta/T$, or more generally for any mode, $x_{(nm)\,max} = \Theta_{nm}/T$.

The specific heat resulting from arbitrary modes of vibration is given by

$$c = 3k_B N^G M_g^{-1} \sum g_{(nm)}(T,\Theta_{nm}) \quad ; \quad J/(gK) \tag{3.7}$$

where N^G is the total number of vibrating masses per mole (Avogadro number) and M_g is the molar mass (see Section 3.7). It is assumed that each vibrating mass is capable of vibrating in three directions of polarization. The summation is made over all modes (nm). Two limiting cases are valid for each mode [3.2; 3.3].

For $T \gg \Theta_{nm}$: $g_{nm} = 1$, and c = const. → (Dulong-Petit law).
For $T \ll \Theta_{nm}$: $g_{nm} \propto T^{n+1}$ and $c \propto T^{n+1}$.

The latter case is important to the specific heat at low temperatures (general Debye law). The temperature dependence of c can thus be determined easily for any mode from the exponent n of the frequency dependence of $D(\omega)$.

$$c(T) \propto T^{n+1}, \text{ if } D(\omega) \propto \omega^n \tag{3.8}$$

Each mode has its own Debye temperature and approaches the Dulong-Petit law at a different temperature. The interchain components c_{3s} and c_{3b} become constant at lower temperatures than the intrachain components c_{1s} and c_{1b}.

The temperature dependences of the most important vibrational modes which contribute to the specific heat of polymers are compiled in Table 3.1. The correlation of the density of states $D(\omega)$, the Debye function g, and the temperature dependence of c are summarized. The relations are valid for the acoustic approximation $T \ll \Theta$.

Table 3.1. Density of states D_i and specific heat c.

Tunneling system	$D_t(\omega) = \text{const.}$	$\to g_t \propto c_t \propto T;$
Three-dimensional continuum	$D_{3s}(\omega) \propto \omega^2$	$\to g_{3s} \propto c_{3s} \propto T^3;\ \Theta_{3s} \approx 80K$
Stiff continuum	$D_{3b}(\omega) \propto \omega^{1/2}$	$\to g_{3b} \propto c_{3b} \propto T^{3/2};\ \Theta_{3b} \approx 128K$
Linear chain	$D_{1s}(\omega) \propto \text{const.}$	$\to g_{1s} \propto c_{1s} \propto T;\ \Theta_{1s} \approx 721K$
Stiff chain	$D_{1b}(\omega) \propto \omega^{-1/2}$	$\to g_{1b} \propto c_{1b} \propto T^{1/2};\ \Theta_{1b} \approx 721K$
Fracton contribution	$D_f(\omega) \propto \omega^{2/3}$	$\to g_f \propto c_f \propto T^{5/3}.$

The Debye temperatures Θ_{nm} of some vibrational modes are included. They are related to HDPE (extrapolated to its fully crystalline state; error ca. 30% [3.2]). It turns out that the acoustic approximation yields rather good results, even at higher temperatures, with $T < \Theta_{nm}$ only (e.g., experimental results prove that three-dimensional bending vibrations obey the relation $c_{3b} \propto T^{3/2}$ up to $T \approx 50K$ despite the fact that $\Theta_{3b} \approx 128K$).

3.3 SPECIFIC HEAT OF POLYMERS

If different vibrational modes are involved, their relative contributions have to be added up. This is shown for crystalline polymers. A distinction is made between contributions from longitudinal and transverse vibrations of aligned polymers. Using the generalized Debye functions $g_{nm}(T,\Theta)$ (Table 3.1) according to Eqs.(3.6) and (3.7), longitudinal and transverse contributions can be added up as follows [3.2]

Longitudinal contributions (stretching phonons)

$$c_s(T) = N_s k_B M_g^{-1} \left[g_{1s}(T,\Theta_{1s}) + \frac{\Theta_{3s}}{\Theta_{1s}} \left\{ g_{3s}(T,\Theta_{3s}) - g_{1s}(T,\Theta_{3s}) \right\} \right] \quad (3.9a)$$

Transverse contributions (mainly bending phonons)

$$c_b(T) = N_b k_B M_g^{-1} \left[g_{1b}(T,\Theta_{1b}) + \left(\frac{\Theta_{3b}}{\Theta_{1b}}\right)^{1/2} \left\{ g_{3b}(T,\Theta_{3b}) - g_{1b}(T,\Theta_{3b}) \right\} + \right.$$
$$\left. + \frac{\Theta_{3w}^{3/2}}{\Theta_{3b}\Theta_{1b}^{1/2}} \left\{ g_{3w}(T,\Theta_{3w}) - g_{3b}(T,\Theta_{3b}) \right\} \cdot \right] \quad (3.9b)$$

The total specific heat is then

$$c(T) = c_s(T) + c_b(T) \quad (3.10)$$

The relative normalizations within longitudinal and transverse vibrations in Eqs. (3.9a,b) are achieved through the respective Debye temperatures Θ_{nm}. Equation (3.9a) constitutes the Tarasov relation, which involves stretching vibrations of a three-dimensional continuum (term g_{3s}), and at higher temperatures only stretching vibrations of a linear chain (term g_{1s}). For the normalization to be accurate, the relation is more complicated. For instance the negative term, $-g_{1s}(T,\Theta_{3s})$, avoids double counting of vibrations. This term has already been incorporated in the term g_{3s}. When taking into account transverse contributions, the situation is even more complicated since not only bending phonons from a three-dimensional continuum (g_{3b}) and a stiff linear chain (g_{1b}) contribute, but also stretching phonons (g_{3w}) are involved. The latter arise from Van der Waals interactions, which influence transverse vibrations in a three-dimensional continuum at very low temperatures.

3.4 EXPERIMENTAL RESULTS ON TEMPERATURE DEPENDENCE

The results from Eqs.(3.9a,b) are compared with experiment in Figure 3.3 for POM (extrapolated to nearly 100 % crystalline fraction) [3.2].

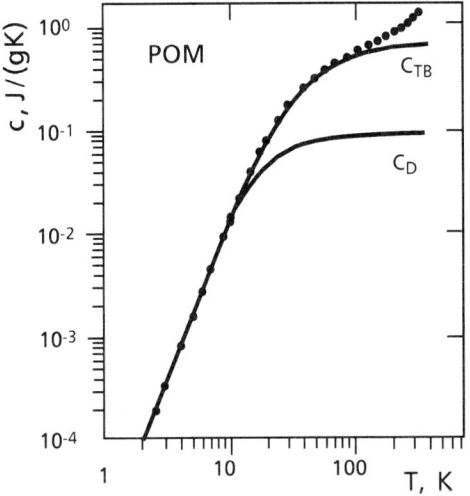

Fig. 3.3. Specific heat of crystalline POM.
c_D: Debye approximation (Eq.3.5a);
c_{TB}: Tarasov-Baur approximation (Eq.3.9).
Experimental results are presented by dots.

The curve c_D has been calculated from Eq.(3.5a). It is valid for an isotropic homogeneous crystal. The curve c_{TB} takes into account the anisotropic microstructure and the bending stiffness of polymers according to the theory of Tarasov and Baur. A very good agreement with experimental results has been found. Equations (3.9a,b), however, contain several insufficiently known parameters. This is especially true for the different Debye temperatures of the individual vibrational contributions. Wunderlich calculated the frequency distribution of many individual vibrational modes in polymers [3.5]. The Debye temperatures were evaluated by application of spectroscopic methods and by measurement of the velocity of phonons. However, the results obtained do not suffice to prove the validity of Eqs.(3.9a,b). They merely indicate that there is a consistency between calculations and measurements.

The next topic treated will be the experimental proof of the calculated temperature dependences of the single modes and the question of which modes are dominant in which temperature ranges. Experiments revealed that there are temperature ranges, where only one mode dominates. In-between, usually only two modes overlap. The treatment is separate for crystalline and amorphous polymers.

SPECIFIC HEAT

3.4.1 Crystalline Polymers

Temperature Range: $T \leq 20K$ (Debye Range)

Down to 0.01K, the Debye behavior $c \propto T^3$ has been found to apply to pure crystals. This is true for Van der Waals crystals (e.g., solid argon [3.6]) or covalently bound crystals (e.g., quartz [3.6]). For polymers the Debye T^3 behavior is also valid, despite the large binding anisotropy. The anisotropy influences the phonon velocity and thus the value of c, but not the temperature dependence. The T^3 law is a consequence of three-dimensional long-wave stretching vibrations (wavelength $\gg$ lattice constant). This behavior is shown in Figure 3.3 for monocrystalline POM and in Figure 3.8 for PE (extrapolated values). The ratio c/T^3, plotted in Figure 3.4a for PE, is constant up to 20K.

Temperature Range: 20K to 60K
(dominant phonon wavelength 3 to 0.5nm)

With increasing temperature, phonons with decreasing wavelengths are excited. The transverse vibrations are more and more influenced by flexural chain stiffness, and bending phonons get active. In addition to the Debye term $c \propto T^3$, a term contributes which results from three-dimensionally propagating bending vibrations

$$c = a_{3s} T^3 + b_{3b} T^m \quad \text{with } m: 3/2 \text{ to } 3 \tag{3.11}$$

The exponent m is continuous since it depends on the propagation direction of bending phonons relative to the axis of the chains. As shown in Section 2.6, $c_{3b} \propto T^{3/2}$ in the direction of the chain only. In all other directions a somewhat higher averaged exponent is valid. The coefficients a_{3s} and b_{3b} characterize the relative contributions from stretching and bending vibrations, respectively. They are slightly dependent on the temperature since in this temperature range the stretching vibrations become saturated while bending vibrations increase. For highly crystalline PE, for example, the term $b_{3b} T^m$ with $m = 3/2$ is dominant in the temperature range above. This is evident from Figure 3.4b where $c/T^{3/2}$ is plotted versus $T^{3/2}$.

Temperature Range: 70K to 200K (Range of Linear Chain Vibrations)
(dominant phonon wavelength $\leq$ 0.5nm)

In this temperature range, the wavelength of energetic vibrations is so short that phonons propagate one-dimensionally in the chain direction. The contribution to the specific heat by longitudinal vibrations of the chain is proportional to T. The transverse vibrational components of stiff chains are proportional to $T^{1/2}$. As a consequence, the total specific heat of flexural, stiff, linear chains is given by

$$c = e_{1s} T + f_{1b} T^{1/2} \tag{3.12}$$

The relative amounts are given by the material specific parameters e_{1s} and f_{1b}. Plotting $c/T^{1/2}$ versus $T^{1/2}$ in Figure 3.4c shows that over the large temperature range of 70K to 200K the results are consistent with a contribution of $c \propto T^{1/2}$. This reveals that the bending phonons are dominant.

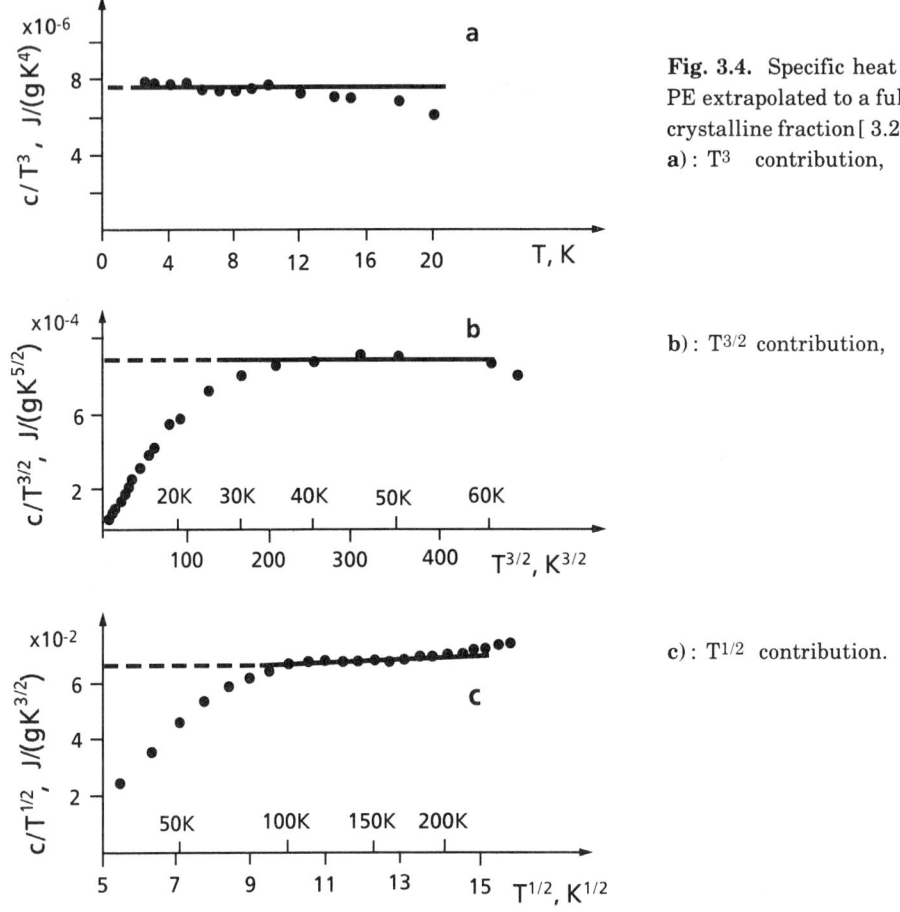

Fig. 3.4. Specific heat of PE extrapolated to a fully crystalline fraction [3.2]:
a): T^3 contribution,
b): $T^{3/2}$ contribution,
c): $T^{1/2}$ contribution.

3.4.2 Amorphous Polymers

For amorphous polymers the temperature dependence is different since additional modes of vibration contribute.

Temperature Range: $T \leq 1K$ (Dominance of Tunneling Vibrations)

At very low temperatures additional contributions to the specific heat are attributed to tunneling vibrations of small molecular units between two neighboring potential minima. The maximum tunneling frequencies are on the order of 10^{12} Hz. The density of states of tunneling systems is assumed to be constant over a wide range of frequencies. From $D_T(\omega) = $ const. it follows from Eq. (3.8) or Table 3.1 that $c \propto T$. Various publications demonstrate the validity of this relation [3.8; 3.9]. An example is shown in Figure 3.5.

For an exact description, the Debye contributions have to be added to the specific heat arising by tunneling vibrations

$$c(T) = g_T T + h_{3s} T^3 \tag{3.13}$$

The parameters g_T and h_{3s} mark their relative contributions. For amorphous polymers the first term $g_T T$ dominates below 1K.

SPECIFIC HEAT

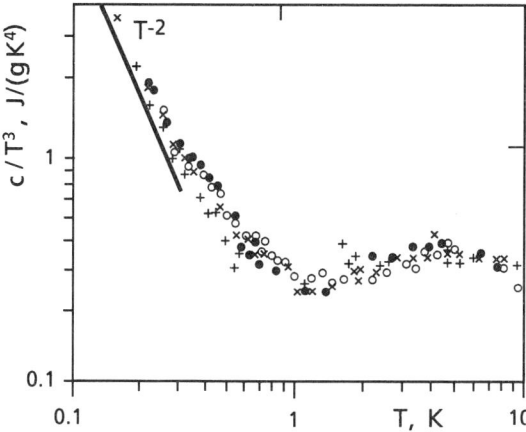

Fig. 3.5. Specific heat of amorphous epoxy resins with different cross-link densities at very low temperatures [3.6]. The range proportional to T^{-2} corresponds to $c \propto T$ in the representation c/T^3.

Temperature Range: 1K to 10K (Debye Range)

For amorphous polymers this range is comparable to the Debye T^3 range of crystals. It turns out that the temperature dependence is even higher for most amorphous polymers. This is attributed to localized vibrations of small molecular units. Their contribution is called "excess specific heat". A temperature dependence stronger than T^3 has, for example, been found for PS and PMMA between 1K and 3K. In Figure 3.6 c/T^3 is plotted versus temperature [3.7; 3.8]. The reason for this strange behavior is still unclear, especially since it is observed at very low temperatures at which a weak temperature dependence from tunneling vibrations becomes dominant.

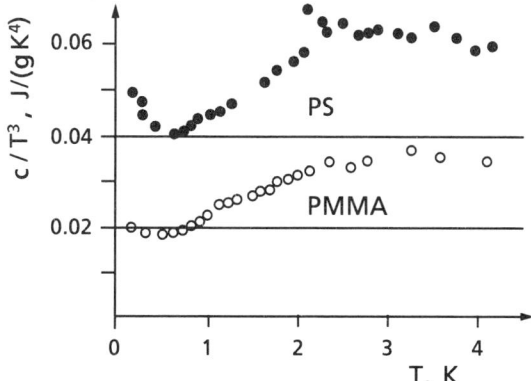

Fig. 3.6. Specific heat of PMMA and PS at very low temperatures. c/T^3 is plotted versus T.

For a series of epoxy resins it has been found that $c \propto T^{3.5}$ between 2K and 8K. The result in Figure 3.7 applies to resins with different cross-link distances $\bar{d}$ and different chemical compositions. It is worth mentioning that specific heats at low temperatures are rather insensitive to the chemical composition and cross-link density of epoxies [3.9].

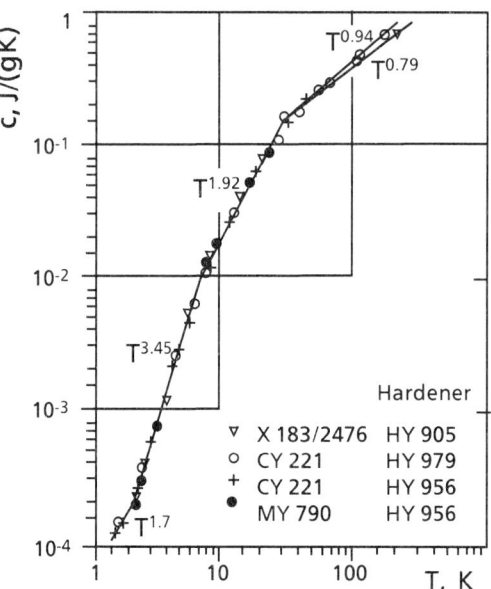

Fig. 3.7. Specific heat of epoxy resins with different cross-link distances versus temperature. Mean cross-link distances $\bar{d}$:1.2 to 12nm.

Temperature Range: 10K to 50K

The influence of bending stiffness of polymeric chains reduces the temperature dependence of specific heat (see Figs.3.7 and 3.8a). It roughly holds: $c \propto T^{1.9}$. This is consistent with the contribution of three-dimensionally propagating bending vibrations $c_{3b} \propto T^m$, with m between 3/2 and 3.

Temperature Range: 50K to 200K (Range of Linear Chain Vibrations)

As seen from Figures 3.7 and 3.12, the specific heat is dominated by contributions from vibrations of linear chains. The temperature dependence is weak, and appears between $T^{0.8}$ and $T^{0.9}$.

In the vicinity of room temperatures and above, contributions from primary glass transition increase the specific heat drastically. As already mentioned, they are superimposed to the basic specific heat, which should be rather independent of temperature in this range (Dulong-Petit law).

3.5 DEPENDENCE ON CRYSTALLINE CONTENT

3.5.1 Specific Heat of Fully Crystalline and Amorphous Polymers

The difference between amorphous and crystalline polymers is demonstrated in Figure 3.8a on PE, whose specific heat has been extrapolated to its completely amorphous and to its completely crystalline phases [3.2]. The specific heat has been measured at different crystalline fractions f_c. According to Figure 3.8b, a linear mixing rule of both phases has been applied. This is a rough procedure, and the curves of Figure 3.8a are not accurate. They show, however, typical trends. Below 50K the specific heat of "amorphous" PE is higher by a factor of 3 to 4. However, no significant difference appears between 80K and 150K. Above 200K, the influence of glass transitions of the amorphous phase becomes dominant.

SPECIFIC HEAT

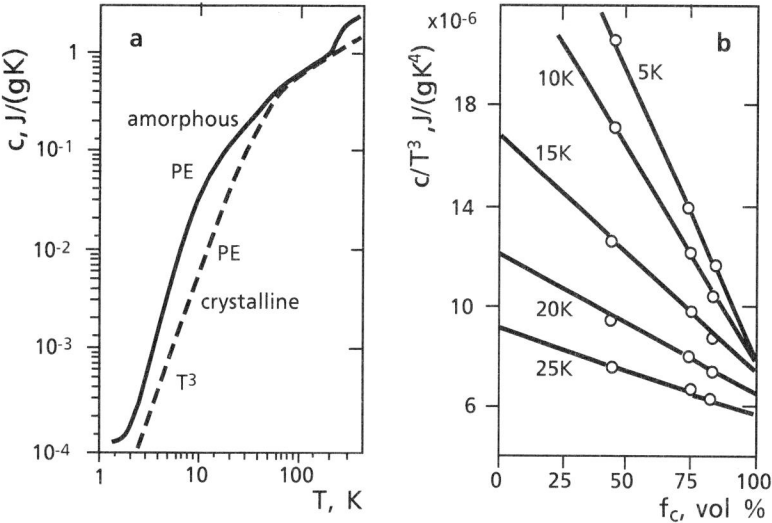

Fig. 3.8. Specific heat of PE. a) dependence on temperature (extrapolated values); b) dependence on crystalline fraction f_c at different temperatures.

The different specific heats of amorphous or crystalline (semicrystalline) polymers can be explained roughly as follows:

Below 50K: $c_{amorph} > c_{cryst}$.

c_{amorph} is higher for the following reasons:

- tunneling contribution below 1K;
- localized vibrations;
- smeared out contribution from weak glass transitions;
- lower mean velocity $\bar{v}$ because of the looser packing of the amorphous phase.

The phonon velocity is powerful since $c \propto 1/\bar{v}^3$ (see Eq.(3.5)). Below 50K phonons propagate in three dimensions, and the mean velocity is smaller for the lower packing density of amorphous polymers.

Between 80K and 150K: $c_{amorph.} \approx c_{cryst}$.

In this temperature range phonon propagation dominates along the chains. The velocity $\bar{v}$ is determined mainly by the chain properties, which are similar for amorphous and crystalline polymers. Crystalline packing is of less importance.

Above 150K:

Glass transitions influence the specific heat of both amorphous polymers and the amorphous phase of semicrystalline polymers, and thus the specific heats are heavily material dependent.

3.5.2 Semicrystalline Polymers

For semicrystalline polymers the dependence of specific heat on the relative crystalline fraction f_c is of interest. An example has already been given for PE in Figure 3.8b. Another is plotted in Figure 3.9 for PETP. Again, in the low-temperature region a dependence on f_c exists, and the amorphous phase ($f_c=0$) exhibits higher values.

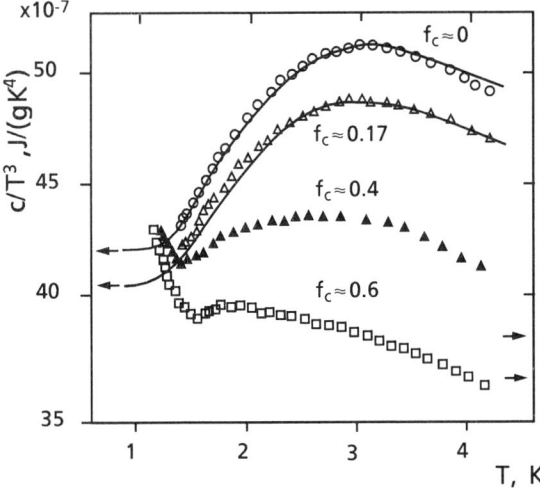

Fig. 3.9. Specific heat of PETP with different crystalline fractions f_c plotted as c/T^3 versus T [3.12].

The conformation of the crystalline domains plays a key role. Lamellar crystallites (PE) have a specific heat different from that of helical structures (PTFE). Looking at Figure 3.8b makes evident that at least between 5K and 25K the specific heat of PE undergoes strong variations by the crystalline fraction f_c. By contrast, for PTFE the specific heat differs only by about 30% when f_c is varied from 0 to 60% at 4K. A variation from $f_c=60$ vol.% to 92 vol.% hardly causes any change of c. The results for PTFE are plotted in Figure 3.10. Both curves with different values of f_c nearly coincide. The curve c_T depicts the Tarasov approximation without taking care of the bending stiffness and optical phonons.

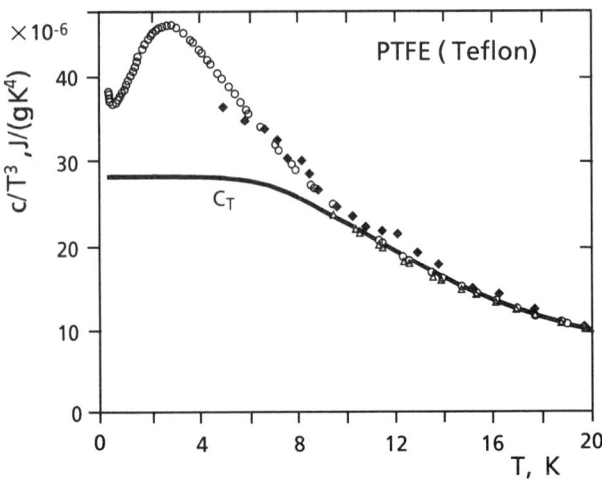

Fig. 3.10. Specific heat of PTFE for different crystalline fractions f_c:
○ : $f_c = 92$ vol.%,
■ : $f_c = 59$ vol.%.
The curve c_T shows the Tarasov's approximation.

A second result is important. The values of c for PTFE are substantially higher than those of PE. Below 10K, they exceed the values which are calculated with the Tarasov theory when using the usual densities of states of a solid and a linear chain. This curve c_T is plotted in Figure 3.10. The following reasons are considered to be responsible for the excessive values of PTFE :

(a) very low phonon velocity;

(b) increased density of states at low frequencies and optical phonons due to the helical structure;

(c) the quasi-crystalline structure of PTFE with amorphous features.

Expanding on each of these reasons:

(a) According to Eq. (3.5) the specific heat is a strong function of the phonon velocity: $c \propto 1/\bar{v}^3$ Due to Eq.(3.5b), mainly the transverse phonon velocity v_t determines $\bar{v}$. According to the relatively high density ρ of PTFE and its smaller modulus G, the phonon velocity $v_t = (G/\rho)^{1/2}$ is below that of many other polymers. For PTFE $v_t = 1.1 \cdot 10^3$ m/s compared to $v_t = 1.8 \cdot 10^3$ m/s for PE. Given a relatively large factor $1/v^3$, this results in a considerably increased specific heat of PTFE.

(b) Moreover, the soft helical structure of PTFE crystallites accounts for the shift of both the acoustic and optical densities of states towards lower frequencies (see Fig.2.9). This increases the specific heat at low temperatures and explains why the specific heat exceeds the curve calculated by the Tarasov approximation. The increased contribution of optical vibrations rises the specific heat. Optical vibrations are determined by the chain properties, which are not different for amorphous and crystalline phases.

(c) One more reason might be the quasi-crystalline structure of PTFE which, while being periodic in the helix direction, is hardly periodic perpendicular to it. Due to the large number of monomers per repetitive unit of the helical structure, periodicity perpendicular to the helix hardly occurs and is not even likely to do so in the crystal-like domains. Given this fact, crystalline PTFE domains exhibit rather the character of aligned polymers without three-dimensional periodicity. Therefore, they exhibit amorphous features, which are indicative of higher specific heat values.

The considerations above make it easier to understand that the specific heat of PTFE is less sensitive to the crystalline fraction and more similar to the behavior of amorphous polymers. Another result is worth mentioning. As shown in Figure 3.11, PTFE behaves more like a crystal if the external pressure is increased [3.12]. At a higher pressure the phonon velocity increases and, consequently, the specific heat decreases. The Debye behavior $c \propto T^3$ becomes more pronounced. As seen from Figure 3.11 the T^3 relation is valid within the temperature range of 1K to 10K at high pressures.

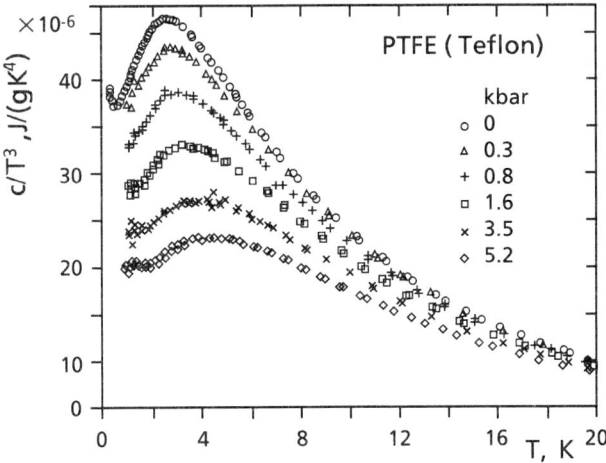

Fig. 3.11. Pressure dependence of the specific heat of PTFE ($f_c = 59$ vol.%).

3.6 GENERAL RESULTS AND DISCUSSION

The experimental results of specific heat have been plotted in Figures 3.12a,b for several amorphous and semicrystalline polymers. Numerical values have been compiled in Table 3A of Appendix 3A [3.14; 3.15]. The following general conclusions can be drawn:

- At low temperatures the specific heat obeys the Debye law and approaches a linear dependence at higher temperatures.

- At low temperatures the specific heat of amorphous polymers exceeds that of semicrystalline ones. An exception is semicrystalline PTFE, whose values lie in the range of amorphous polymers.

- The specific heat of most amorphous polymers does not depend very much on the chemical composition. A typical exception is PS which below 50K exhibits excessive values as shown in Figure 3.12. The reason might be that contributions from its tertiary glass transition near 40K are added.

- The specific heat of semicrystalline polymers is not very sensitive to the chemical structure at low temperatures. As already mentioned, PTFE exhibits relatively high values. Its behavior is similar to that of an amorphous polymer, and one striking feature is its insensitivity to the crystalline fraction.

- In the range of 80K to 100K the specific heats of amorphous and semicrystalline polymers are similar.

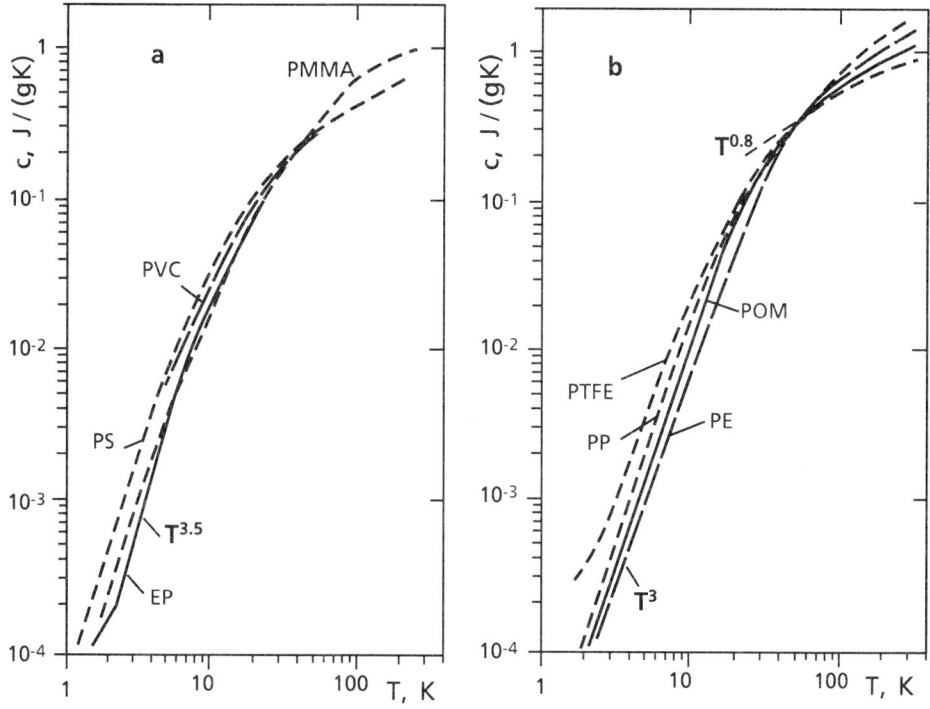

Fig. 3.13. Specific heat of polymers. a) amorphous; b) semicrystalline

3.7 MOLAR HEAT CONVERSION

The conversion of the mole related specific heat, c_M, to the mass related value c is given by

$$c\,(J/(g\ K)) = M_g^{-1} c_M\,(J/(mole\ K))$$

M_g is the molecular weight of one repetitive unit (e.g., $M_g = 28$ g/mole for a repetitive unit CH_2-CH_2 of PE). For molecular conversion equations, see Ref [3.13]. Typical values are shown in Table 3.2.

Table 3.2. Mole conversion.

Polymer	M_g, g/mole
HDPE	28
PS	104
PVC	62.2
PMMA	100
PTFE	100
POM	30

3.8 SUMMARY

(1) The specific heat of polymers is characterized by a three-dimensional phonon propagation at low temperatures and by a one-dimensional propagation along the molecular chains at higher temperatures (Tarasov theory).

(2) The bending stiffness of polymer chains has to be taken into account for transverse components of vibrations. Bending vibrations become dominant at elevated temperatures.

(3) The temperature dependence of specific heat is determined by the frequency dependence of the density of states $D(\omega)$ of a vibrational mode. Each mode is characterized by a specific exponent n of a power law ω^n.

(4) The general relation between the temperature dependence of the specific heat, c(T), and the frequency dependence of the density of states, $D(\omega)$, is given by $c(T) \propto T^{n+1}$, if $D(\omega) \propto \omega^n$. This relation is valid for the acoustic approximation, $T \ll \theta$, for all exponents n. The specific heat is rather well described by the acoustic approximation. At several temperature intervals different vibrational modes are superimposed.

(5) The vibrations of each mode i get saturated at a specific temperature, the Debye temperature Θ_i.

(6) Amorphous polymers show a higher specific heat than semicrystalline ones. The phonon velocity which determines c is smaller for the loosely packed amorphous phase than for crystalline ones. (PTFE, by contrast, is an example, which shows a rather high specific heat and little dependence on the crystalline content; the reason is the soft helix structure of the crystallites and contributions from optical phonons.) Furthermore, at low temperatures, localized vibrations and tunneling vibrations contribute to the specific heat of amorphous polymers. The temperature dependences are, therefore, different for amorphous and crystalline polymers.

(7) In most cases, the specific heat of polymers is more sensitive to the phases (amorphous or semicrystalline) than to the chemical composition. Below 100K, the specific heat is not strongly dependent on the chemical composition, when related to the same crystalline fraction (see Table 3A in the Appendix 3A).

3.9 REFERENCES

3.1 Kittel, C., Introduction to Solid State Physics, p.241, John Wiley and Sons (1973).

3.2 Baur, H.; Colloid Z. u. Z. Polym. 241, p.1057, (1970),
ibid. 244, p. 293, (1971),
ibid. 250, p. 1015, (1972).

3.3 Engeln, J., doctoral thesis, TU-Berlin 1983, D83; and Engeln, J. and Meißner, M., in: "Nonmetallic Materials and Composites at Low Temperatures",Vol. 2; p.1 Eds: Hartwig, G., Evans, D., Plenum Press, N.Y. and London, (1982).

3.4 Engeln, J. and Wolbig, D.; Coll. Polym. Sci. 261 (1983); p.736.

3.5 Wunderlich, B. and Baur, H.; Adv. Polym. Sci. 7 (1970); p.151 and Kirkpatrik, D.E.; Judovits, L. and Wunderlich, B.;J. Polym. Sci.; Phys. Ed. 24 (1986); p.45.

3.6 Kittel, C.: Introduction to Solid State Physics; p. 262; John Wiley Publ. (1973).

3.7 Gaur, U.;and Wunderlich, B. ;J. of Phys. and Chem. Reference Data; Vol. 11 (1982) and Vol. 12 (1983).

3.8   Grebowicz, J.; Lau, S.F. and Wunderlich, B.;J. Polym. Sci. 71 (1984); p.17.
3.9   Hartwig, G.; Prog. in Colloid. and Polym. Sci. 64 (1984); p.56.
3.10 Nicholls, C.J. and Rosenberg, H.M.; J. Phys. C., 17 (; p.1165 andPhysica 108 B (1981); p.1015.
3.11 Hunklinger, S. and Raychaudhuri, A.K.; in Proc. Low Temp. Phys. 9 (1986); p.265 Ed. T. F. Brewer, North Holland Publ. Comp.
3.12 Boyer,J.D.; J.C. Lasjaunias; R.A. Fisher and N.E. Phillips; J. Non-Cryst. Sol. Vol.55; (1983); p. 423.
3.13 Krevelin, D.W.; P.J. Hoftyzer: Properties of Polymers; Elsevier Publishing Comp.; (1972); p. 44.

THERMAL EXPANSION AND THE GRUENEISEN RELATION

Contents

4	Thermal expansion and the Grueneisen relation	69
4.1	Definition and survey	69
4.2	Basic thermal expansion and vibrational modes	72
4.3	Thermal expansion from glass transitions	80
4.4	Thermal expansion and cross-linking	83
4.5	Grueneisen relation	84
4.6	Temperature dependence of the thermal expansion and of the Grueneisen parameter	88
4.7	Summary	93
4.8	References	95

Appendix

4A   Data

4 THERMAL EXPANSION AND THE GRUENEISEN RELATION

4.1 DEFINITION AND SURVEY

4.4.1 Thermal Expansion

The linear thermal expansion coefficient a is defined by the relative length change per temperature variation

$$a = \frac{dL}{dT}\frac{1}{L} \quad ; K^{-1} \tag{4.1a}$$

For an isotropic solid the volume expansion coefficient β is given by

$$\beta = \frac{dV}{dT}\frac{1}{V} = 3a \quad ; K^{-1} \tag{4.1b}$$

The experimentally measured quantity is the integral thermal expansion

$$\frac{\Delta L}{L_0} = \int_{293}^{T} a(T)\, dT \tag{4.2}$$

where the reference temperature is 293K and $L_0 = L(293K)$. In this definition, for a positive a, the quantity $\Delta L/L_0$ becomes negative for cooling. For many technical purposes the integral thermal expansion is more directly applicable, but for physical interpretations the differential quantity a is of more significance. Thermal expansion is one of the thermodynamic properties which has been investigated for three centuries. Nevertheless, theoretical explanations are still difficult, since several types of binding potentials and vibrational modes are involved and influences from the glass transitions are superimposed.

Thermal expansion occurs if there are anharmonic interactions between the vibrating particles. The vibrational amplitudes of a mode as a function of temperature determine the expansion behavior. The amplitudes are expressed by the density of phonons which are excited at a temperature T.

Due to the chain structure of polymers different vibrational modes can be activated. Thermal expansion caused by stretching vibrations arises from their anharmonic components which are governed by the asymmetry of the binding potentials. Intra- and interchain stretching vibrations are controlled by the less asymmetric covalent potential and by the more asymmetric Van der Waals potential, respectively. Thus, interchain vibrations yield a larger thermal expansion. Thermal expansion arising from stretching vibrations is always positive.

The bending (transverse) vibrations exhibit a different behavior. As the temperature increases, the amplitudes of transverse vibrations get larger and the time average of the chainlength gets smaller. This results in a negative component to the thermal expansion in the chain direction. Isotropic amorphous polymers are entangled and all contributions are combined. A schematic representation of the temperature dependence of the thermal expansion is given in Figure 4.1.

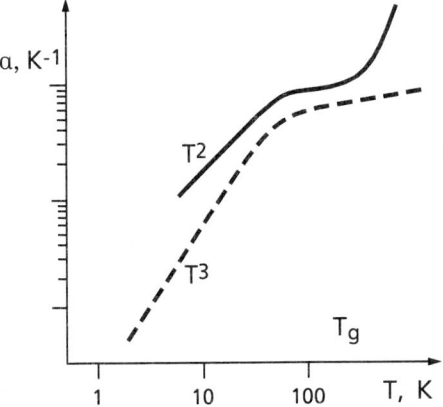

Fig.4.1. Coefficient of thermal expansion shown schematically versus temperature.
——— amorphous
- - - crystalline

A strong temperature dependence has been found below about 70K. In this range three-dimensional interchain vibrations are excited whose density increases strongly with temperature (see Section 3.3). A T^3 dependence has been found for highly crystalline polymers. Amorphous polymers show a weaker temperature dependence. Above 70K thermal expansion of polymers generally is rather weakly dependent on temperature. The reason is that interchain vibrations are more or less saturated, and linear intrachain vibrations are superimposed whose density increases little with temperature. At elevated temperatures, however, the thermal expansion is strongly increased by glass transitions at which the temperature dependence of the free volume changes. The contributions from glass transitions are superimposed on the contributions from thermal vibrations (basic thermal expansion). Thermal expansion of amorphous polymers above 100K arises mainly from the glass transitions. Crystalline polymers, however, exhibit no secondary or tertiary glass transitions and exhibit therefore a much smaller thermal expansion. For semicrystalline polymers the expansion behavior is determined mainly by the amorphous phase.

4.1.2 Grueneisen Relation

Some similarity of the thermal expansion to the behavior of the specific heat is seen (compare Figs. 3.1 and 4.1). This similarity can be understood by the Grueneisen relation which states that the coefficient of thermal expansion a is proportional to the specific heat c. The Grueneisen relation is given by

$$a = \frac{1}{3} \frac{\rho}{K} \gamma \, c \tag{4.3}$$

where ρ is the density, K the bulk modulus (inverse compressibility) and γ is the Grueneisen parameter. The specific heat is a measure of the density variation of vibrations with temperature and the Grueneisen parameter is a measure of the expansion capability of the vibrations. For stretching vibrations, for example, the Grueneisen parameter is determined by the asymmetry of the binding potential between vibrating units. It is larger for interchain stretching vibrations than for intrachain ones. For crystalline polymers γ_{tot} is larger at low temperatures due to interchain vibrations γ_{inter}. At higher temperatures, intrachain vibrations with a low γ dominate γ_{intra}. The Grueneisen parameter of amorphous polymers starts at low γ_T due to tunneling vibrations, runs through a high maximum and drops down to the low value of γ_{intra}. For aligned polymers the anisotropic behavior is described by a Grueneisen tensor (see Section 4.6.3).

From Eq.(4.3) it follows that for crystalline polymers below 20K, the T^3 law of the specific heat is valid also for the thermal expansion coefficient if γ is constant. The Grueneisen relation qualitatively explains the higher thermal expansion coefficient a of amorphous polymers since their specific heat c is larger and the bulk modulus K is usually smaller than for crystalline phases. The remaining question is if there is a temperature dependence of the Grueneisen parameter γ. If the bulk modulus K is constant and only one vibrational mode determines both a and c, then γ is constant and independent of temperature. Generally, however, several vibrational modes n contribute to a total Grueneisen parameter with different γ_n. Since most vibrational modes are dominant at different temperature ranges, the total Grueneisen parameter γ_{tot} is a strong function of temperature and is different for amorphous and crystalline polymers at low temperatures. At glass transitions the bulk modulus K drops and a rises. It will be shown later on that γ is rather insensitive to glass transitions because the product $a \cdot K$ stays rather constant. A schematic representation of this behavior is given in Figure 4.2 (see Section 4.6.3).

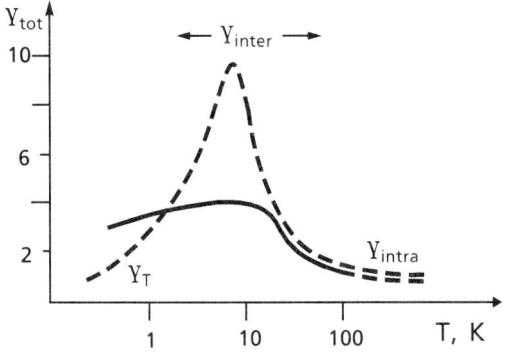

Fig.4.2. The total Grueneisen parameter γ_{tot} as a function of temperature (schematic).
—— crystalline
--- amorphous

The microscopic definition of the Grueneisen parameter is given by the relative frequency shift per relative volume change.

$$\gamma = - \frac{d\omega}{\omega} / \frac{dV}{V} = - d\ln \omega / d\ln V \qquad (4.4)$$

The volume change (e.g., by external pressure) causes an increase of the internal "thermal pressure," which is manifested by a frequency shift. A constant value of γ means that the relative frequency shift is independent of frequency. The Equation(4.4) reflects the basic assumption neccessary for deriving of Eq.(4.3) (see Subsection 4.6.1). Since the bulk modulus K (inverse compressibility) is involved in the Grueneisen relation, γ can be used for evaluating the pressure dependence of several thermal and mechanical properties (Eqs.(4.25) to (4.27).)

4.2 BASIC THERMAL EXPANSION AND VIBRATIONAL MODES

The so-called basic thermal expansion arises from thermal vibrations and comprises that portion which is not influenced by glass transitions. The thermal expansion behavior is similar to that of the specific heat, as predicted by the Grueneisen relation. The temperature dependence of the basic thermal expansion depends on the density of vibrations (phonons) as a function of temperature. A strong temperature dependence exists below about 70K. In this range three-dimensional vibrations are excited whose density increases strongly with temperature (see Section 3.3). Since mainly interchain vibrations are involved, the increase of thermal expansion is rather large. Above 70K thermal expansion is rather weakly dependent on temperature. Interchain vibrations are more or less saturated, and linear, isolated intrachain vibrations are superimposed whose density increases little with temperature. Thermal expansion from intrachain vibrations, moreover, is relatively small. This is owing to the less asymmetric covalent binding potential and owing to transverse vibrations which contribute negative components of thermal expansion. Thus, basic thermal expansion of polymers increases weakly with temperature above about 70K. The basic thermal expansion can be studied for example on amorphous PS, which shows no glass transition below room temperature, except a small tertiary one around 40K. As seen in Figure 4.3, there is only a slight temperature dependence of a above 60K. Other examples will be shown in Figure 4.8 for crystalline POM and in Figure 4.9 for LCPs.

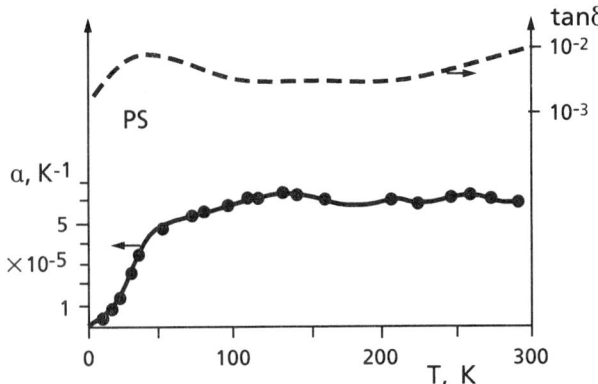

Fig. 4.3. Thermal expansion of amorphous PS. For comparison the damping spectrum tanδ at 10Hz is added.

4.2.1 Isotropic Thermal Expansion

Thermal expansion of polymers depends strongly on the modes of vibration and the binding potentials involved. The entanglement of chains is a further complication. A rough approximation can be applied which calculates the isotropic thermal expansion coefficient a_{iso} from the longitudinal and transverse components $a_\parallel$ and $a_\perp$ of completely aligned polymers.

$$a_{iso} \approx \frac{1}{3}(a_\parallel + 2a_\perp) \qquad (4.5)$$

These components are more accessible to calculations. The approximation is made that the expansion $a_\perp$ in both directions perpendicular to the chain is equal. Different vibrational modes along and perpendicular to the molecular chains yield an anisotropic expansion, which can be observed on aligned polymers. The density of vibrations is a measure of the vibrational amplitudes.

4.2.2 Longitudinal and Transverse Components

A model of fully aligned polymers is assumed for discussion. The following vibrational modes contribute to $a_\parallel$ and $a_\perp$, respectively (see Figs. 4.4 and 4.6):

Components of $a_\parallel$

$a_{\parallel s}$: longitudinal intrachain stretching vibrations of atoms in the covalent potential.

$a_{\parallel b}$: transverse (bending) vibrations and their projection onto the axis.

Components of $a_\perp$

$a_{\perp sw}$: transverse interchain stretching vibrations of segment axes in the Van der Waals potential (dominant below 10K).

$a_{\perp b}$: transverse (bending) vibrations relative to the chain axis, mainly driven by intrinsic bending stiffness (dominant above 10K).

In the chain direction there are contributions from longitudinal and transverse vibrations. Perpendicular to the chain a distinction is made between $a_{\perp sw}$ and $a_{\perp b}$. The term $a_{\perp sw}$ considers stretching vibrations of large segments of neighboring chains, which can be calculated from the asymmetry of the Van der Waals potential. Their vibrating amplitudes are controlled by the Van der Waals potential only. The term $a_{\perp b}$ takes into account that components of the amplitudes which are mainly controlled by the intrinsic bending stiffness of chains. This influence becomes dominant at elevated temperatures. The common feature of the components $a_{\parallel s}$, $a_{\perp sw}$ and $a_{\perp b}$ is that vibrational amplitudes increase with temperature and cause a positive coefficient of thermal expansion. The transverse vibration term $a_{\parallel b}$ is of different nature. It causes a negative coefficient of thermal expansion. The following sections will give a brief treatment of the thermal processes which control thermal expansion. Additional information is given by the Grueneisen relation, which will be dealt with in Section 4.5.

Thermal Expansion from Stretching Vibrations. Stretching vibrations are governed by the separating distance of the vibrating units. If the binding potential between vibrating units is asymmetric, anharmonic oscillations result. A change of their amplitudes leads to a change of their mean distance, averaged over time. This is shown schematically by dashed lines in Figure 4.4 for the covalent and the Van der Waals potentials. If the temperature is increased, higher energy levels are excited, and the mean distance between oscillators gets larger because of the stronger repulsive forces. In the vicinity of their minima the binding potentials, however, are symmetric. Since at low temperatures only low-energy levels are excited, the thermal expansion goes to zero at $T \to 0$.

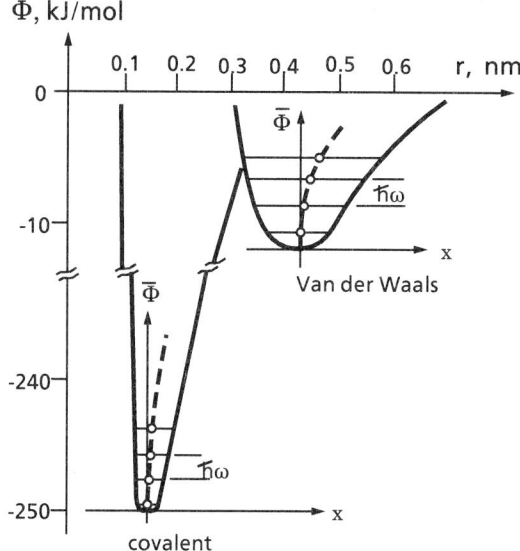

Fig. 4.4. Covalent and Van der Waals potentials as a function of the relative distances r between oscillating units.

Longitudinal intrachain vibrations are controlled by a covalent potential, which has only a small asymmetry; the thermal expansion $a_{\| s}$ is thus small (Fig. 4.5a).

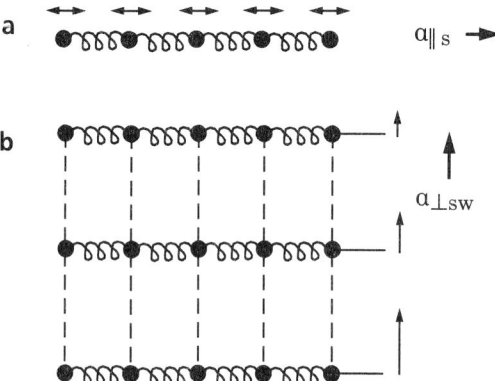

Fig. 4.5. Stretching vibrational modes. a) atoms vibrating in the covalent potential. b) chain axes vibrating in the Van der Waals potential (dashed lines).

Interchain vibrations of the chains or chain segments are controlled by their mutual distance (see Fig.4.5b) and are driven by the Van der Waals potential which is rather asymmetric. $a_{\perp sw}$ is therefore relatively large. The term $a_{\perp b}$ is not determined by the asymmetry of the potentials but by the chain stiffness, which induces a weaker temperature dependence of the vibrational amplitudes.

The bending gets larger with decreased wavelengths. Its influence starts at about 15K.

Theoretical Background. The asymmetric binding potential Φ can be expanded in a power series of the distance x whose origin is put at the potential minimum (see Fig.4.4).

$$\overline{\Phi}(x) = sx^2 - ax^3 + \cdots \qquad (4.6a)$$

The odd term ax^3 and the higher order terms are decisive for the thermal expansion, and s and a are constants. The temperature dependence of the mean distance $\bar{x}$ can be obtained by applying Boltzmann statistics [4.1].

$$\bar{x} = \frac{\int x \exp(-\overline{\Phi}/k_B T) \, dx}{\int \exp(-\overline{\Phi}/k_B T) \, dx} = \frac{3k_B a}{4s^2} T \qquad (4.6b)$$

According to Eq.(4.1a) the thermal expansion from stretching vibrations is

$$\alpha_s = \frac{dx}{dT}\frac{1}{r_0} = \frac{3k_B a}{4s^2 r_0} \qquad (4.6c)$$

where r_0 is the mean equilibrium distance. This calculation yields a constant thermal expansion, which is valid only over a small range at high temperatures. The Boltzmann statistics deal with oscillators which have continuous energies equal to $k_B T$. At lower temperatures the quantum nature of phonons becomes dominant and can be properly described by Bose-Einstein statistics. The thermal expansivity then becomes a function of temperature. This method for calculating α_s, by means of an asymmetric binding potential, has been applied in several publications [4.2 to 4.4]. In principle the procedure is similar to the calculation of specific heat. This includes the application of density of states for all vibrational modes involved and the probability of states. But for calculating thermal expansion an additional parameter is necessary which describes the anharmonicity of vibrations. This parameter is given by the Grueneisen parameter (see Section 4.4). It can be shown that below $\approx 15K$ for crystalline polymers $\alpha \propto T^3$ [4.5].

Thermal Expansion from Bending Vibrations. The large free volume between chains gives space for transverse vibrations. A model of an aligned polymeric chain with length L_λ is applied (see Fig.4.6). An increase in temperature results in larger transverse amplitudes A, and thus in a reduction of the mean time-averaged chainlength. This results in a negative thermal expansion coefficient in the chain direction. The thermal expansion arising from transverse vibrations is mainly of geometrical nature and is determined by the projection onto the chain axis. L_λ is the stretched length of a wavelength piece.

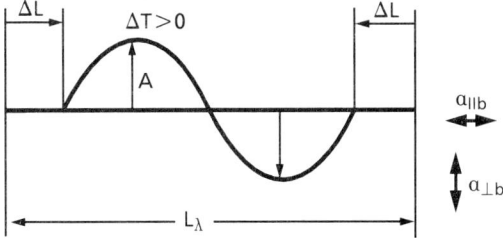

Fig. 4.6. Transverse vibrations of aligned chains with a transverse amplitude A.

Theoretical Background. A rough calculation of the component $a_{\|b}$ from transverse vibrations in the chain direction gives

$$a_{\|b}(\omega,T) \approx -\frac{2A}{\lambda^2}\frac{dA}{dT} \approx -2A\left(\frac{\omega}{2\pi v}\right)^2 \frac{dA}{dT} \tag{4.7a}$$

where v is the phonon velocity and ω the phonon frequency. It is assumed that the chainlength is much larger than the wavelength λ. The thermal contraction $a_{\|b}$ increases with the transverse amplitude A and quadratically with the frequency ω. Since dA/dT is positive, a negative thermal expansion exists in the chain direction. The strong dependence on ω (or λ) is clear, since at high frequencies many wavelength pieces L_λ per total chainlength add up to cause a larger negative $a_{\|b}$. The frequency range, however, is limited by the maximum frequency ω_{1bmax} which has been introduced for calculating specific heat. Qualitatively, the amplitude A is related to the thermal energy U_b from bending vibrations. It is proportional to the squared amplitude: $U_b \propto A^2$. This yields

$$\frac{dA}{dT} \propto \frac{1}{2A}\frac{dU_b}{dT} = \frac{1}{2A}c_b(T) \tag{4.7b}$$

$$a_{\|b}(\omega,T) \propto -\left(\frac{\omega}{2\pi v}\right)^2 c_b(T) \quad \text{for } \omega \leq \omega_{b\,max} \tag{4.7c}$$

where $c_b(T)$ is the specific heat contributed from bending vibrations (see Section 3.3). Since $\omega \propto T$ it is obvious that negative thermal expansion $a_{\|b}$ increases strongly with temperature. At low temperatures is $c_b \propto T^{3/2}$, and therefore $a_{\|b} \propto -T^{7/2}$. This is roughly consistent with a T^3 dependence. Thus, at low temperatures the thermal expansion caused by stretching and bending vibrations exhibits nearly the same temperature dependence; the thermal expansion coefficient, however, is of opposite sign. Results are shown in Figure 4.8 on monocrystalline POM. The components $a_\|$ in the chain direction are dominated by the bending component $a_{\|b}$, which shows a negative thermal expansion coefficient and a T^3 dependence. The transverse thermal expansion $a_\perp$ is positive and obeys the same temperature dependence.

Total Thermal Expansion caused by Vibrations. The total thermal expansion components $a_{\|\,tot}$ and $a_{\perp\,tot}$ are a combination of contributions from longitudinal and transverse vibrations as shown in Figure 4.7.

$$a_\| \equiv a_{\|\,tot} \approx a_{\|s} + a_{\|b} \tag{4.8a}$$

$$a_\perp \equiv a_{\perp\,tot} \approx a_{\perp sw} + a_{\perp b} \tag{4.8b}$$

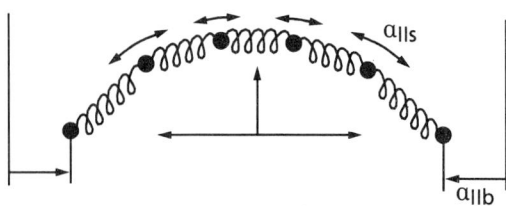

Fig. 4.7. Thermal expansion behavior resulting from longitudinal and transverse vibrations.

For many polymers the negative transverse component $a_{\|b}$ is dominant and the value of $a_{\|tot}$ is slightly negative. A well-known example are Kevlar fibers, whose aligned aramid molecules show a negative $a_\|$.

4.2.3 Results on Fully Aligned Polymers

Monocrystalline polymers, pure liquid crystalline polymers (LCPs) and highly stretched polymers exhibit full alignment. For monocrystalline POM the thermal expansion components $a_\|$ and $a_\perp$ are plotted in Figure 4.8. Below 20K both, $a_\|$ and $a_\perp$ obey a T^3 law, where $a_\|$ is small but negative.

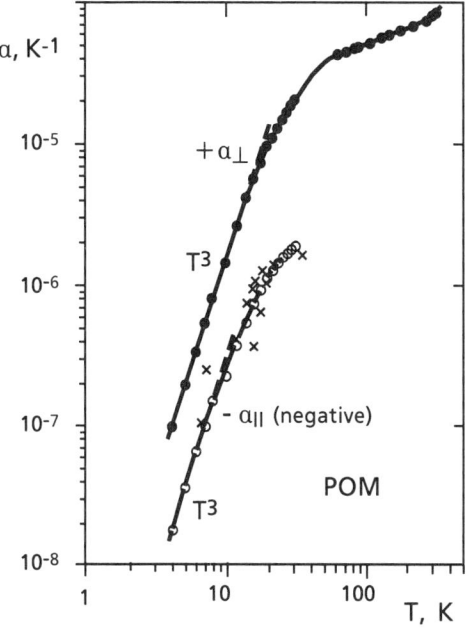

Fig. 4.8. Thermal expansion $a_\|$ (negative) and $a_\perp$ of monocrystalline POM versus temperature [4.5; 4.6].

LCPs are self-orienting polymers, which can be processed to have a rather good orientation after polymerization. The values $a_\perp$ and $a_\|$ for Vectra A 950 are plotted in Figure 4.9. As expected $a_\|$ is negative.

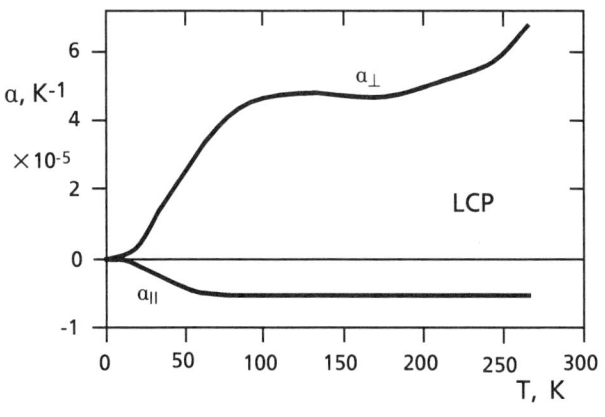

Fig. 4.9. Anisotropic thermal expansion of LCP (Vectra A 950).

An example of thermal expansion of aligned HDPE is given in Figure 4.10a. For comparison the values a_{iso} of isotropic HDPE are added (see Eq.4.5). As expected, $a_\parallel$ is negative and $a_\perp$ is positive and larger than a_{iso}.

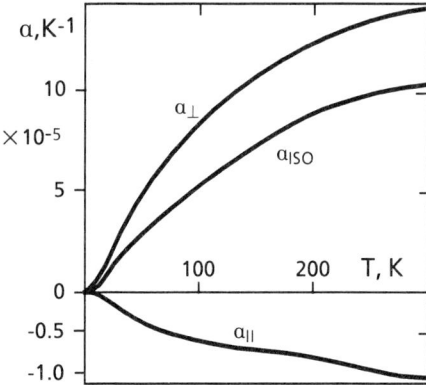
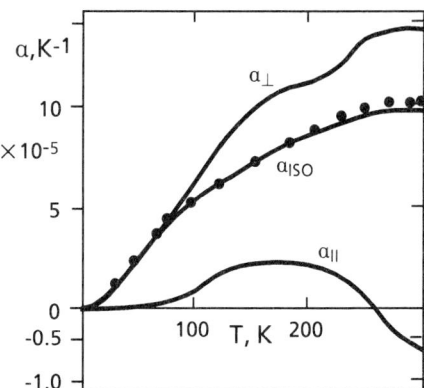

Fig. 4.10a. Thermal expansion of isotropic and fully aligned HDPE [4.7; 4.8].

Fig. 4.10b. Thermal expansion of isotropic and 200 % stretched HDPE. Dots from Eq. (4.5).

For comparison, the results on not fully aligned HDPE are shown in Figure 4.10b. Negative thermal expansion of $a_\parallel$ starts first at high temperatures as expected from Eq.(4.7c); (see further Ref.4.25 and 4.26).

4.2.4 Thermal Expansion of Aligned Polymers in Off-Axis Directions

The directions parallel and perpendicular to the chain alignment are called orthotropic directions. The thermal expansion in a direction θ relative to that of the alignment can be calculated from the orthotropic values $a_\parallel$ and $a_\perp$ by applying an angle transformation matrix T (see Eq.11.9).

$$\begin{bmatrix} a_x \\ a_y \\ \Gamma_{xy} \end{bmatrix}_\theta = T \cdot \begin{bmatrix} a_\parallel \\ a_\perp \\ 0 \end{bmatrix}_{ortho} \tag{4.9}$$

It can be shown that there exists a third component Γ_{xy}, if $a_\perp \neq a_\parallel$. This component is called the coefficient of "thermal shear expansion". It is defined by the thermal shear angle δ caused by a temperature variation ΔT.

$$\Gamma_{xy} = -\delta/\Delta T \tag{4.10a}$$

Evaluation of Eq.(4.9) yields (see Section 11.2)

$$\Gamma_{xy} = \left(a_\perp - a_\parallel\right) \sin 2\theta \tag{4.10b}$$

The thermal shear deformation is plotted in Figure 4.11. A rectangular piece, cut in an off-axis direction, deforms to a rhombic shape with temperature variations. The cause is a very different thermal expansion parallel and perpendicular to the chain orientation. In the chain direction $a_\parallel$ is small and perpendicular $a_\perp$ is large.

Any substance with an anisotropic expansion tensor will exhibit a thermal shear deformation in the off-axis directions.

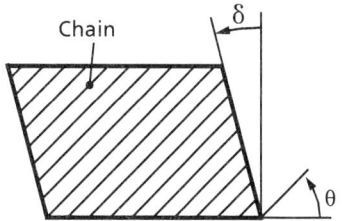

Fig. 4.11. Thermal deformation of a rectangular piece of aligned polymers in an off-axis direction θ for $\Delta T > 0$.

4.2.5 Thermal Expansion of Not Fully Aligned Polymers

The components of thermal expansion for a not fully aligned polymer are shown in Figure 4.10b. They differ from those of the fully aligned case at which $a_\parallel$ is negative only at rather high temperatures (Fig. 4.10a). The negative term $a_{\parallel b}$ increases strongly with frequency ω, and dominates first at high temperatures (see Eq.(4.7c)). The thermal expansion of not fully aligned polymers can be calculated by introduction of an orientation function f.

$$f = 0.5 \left(\overline{3 \cos^2 \theta} - 1 \right) \tag{4.11a}$$

where θ is the orientation of a chain element relative to the drawing direction [4.9; 4.10]. For isotropic polymers $f = 0$ and for fully aligned ones $f = 1$. Assuming small chain elements with thermal expansions $a_\parallel^u$ and $a_\perp^u$ one can calculate the resulting thermal expansion of partially aligned polymer chains:

$$a_\parallel = 1/3 \left[(1+2f) a_\parallel^u + 2(1-f) a_\perp^u \right] \tag{4.11b}$$

$$a_\perp = 1/3 \left[(1-2f) a_\parallel^u + (2+f) a_\perp^u \right] \tag{4.11c}$$

In most cases it is difficult to calculate f. An experimental determination can be obtained from measuring $a_\parallel$, $a_\perp$ and a_{iso} (assuming that $f=0$). If $a_\perp^u \gg a_\parallel^u$ it holds for f being well below 1:

$$a_\parallel \approx \frac{2}{3} (1-f) a_\perp^u \tag{4.12a}$$

$$a_\perp \approx \frac{1}{3} (2+f) a_\perp^u \tag{4.12b}$$

$$f \approx 2 \frac{a_\perp - a_\parallel}{a_\parallel + 2 a_\perp} \approx \frac{2}{3} \frac{a_\perp - a_\parallel}{a_{iso}} \tag{4.13}$$

The last term of Eq.(4.13) is obtained from Eq.(4.5). If the value of f is determined from Eq.(4.13) the mean orientation θ can be calculated with Eq.(4.11a).

4.3 THERMAL EXPANSION FROM GLASS TRANSITIONS

4.3.1 Survey

As already mentioned thermal expansion is not only determined by vibrational modes but is also influenced to a large degree by glass transitions. Here, the temperature dependence of the free volume changes and with it the thermal expansion. This is especially true for the loosely packed molecular chains in amorphous polymers. Thermal expansion can be considered as a thermally induced vibrational pressure against binding forces. Hence, when the volume available for the internal pressure of the vibrations is increased, the expansivity rises [4.11]. At low temperatures the vibrating mass is locked in a potential minimum (see Fig. 4.12). In the range of a transition temperature the relevant segments or side groups start unfreezing thus increasing the free volume and the thermal expansion. Unfreezing in this respect means that the maximum vibrational energy within a thermal energy distribution exceeds the potential barrier between neighboring potential minima. Hence, a larger volume becomes available for vibrations and the internal pressure.

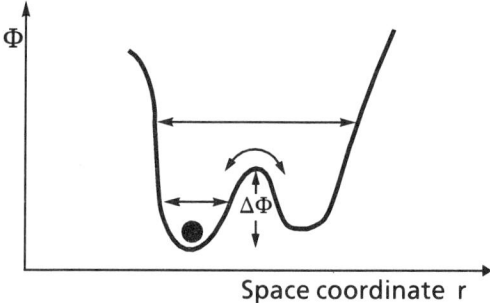

Fig. 4.12. Double potential well with a barrier $\Delta\Phi$. The double arrows symbolize vibrations at different energies (or temperatures).

At primary transitions the effect of the free volume change is much larger than for secondary or tertiary ones. For primary glass transitions collective place changes of several segments are involved which yield a rather sharp transition, while at secondary or tertiary transitions only small molecular units become mobile. For primary glass transitions Simha and Boyer derived a universal relation for the change of the expansion by volume, $\Delta\beta$, between temperatures just below and above T_g [4.11]. The extra free volume is characterized by g.

$$\Delta\beta \, T_g = g = 0.113 \qquad (4.14)$$

This relation cannot be transfered directly to secondary glass transitions, since their volume change is a smoother function of temperature. However, the influence of secondary or tertiary transitions on the thermal expansion is similar. There is a superposition of the basic expansion from vibrations and the contributions from glass transitions as shown schematically in Figure 4.13. As already mentioned, the basic thermal expansion above $\approx 70K$ increases only slightly with temperatures. At glass transitions, however, a drastic increase of a can take place.

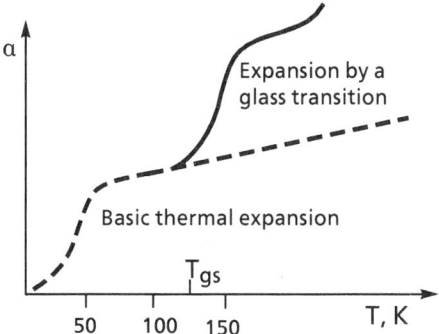

Fig. 4.13. Superposition of basic thermal expansion and a contribution from a secondary glass transition at T_{gs}.

4.3.2 Thermal Expansion Results in the Regime of Glass Transitions

A great difference exists between amorphous and crystalline polymers with respect to their glass transitions. An example is shown in Figure 4.14 for PE, extrapolated to its fully crystalline and fully amorphous phases. In the temperature range shown in this figure, no glass transition of the crystalline phase occurs; the thermal expansion above 70K rises only slightly with temperature, and reflects the behavior of the basic thermal expansion.

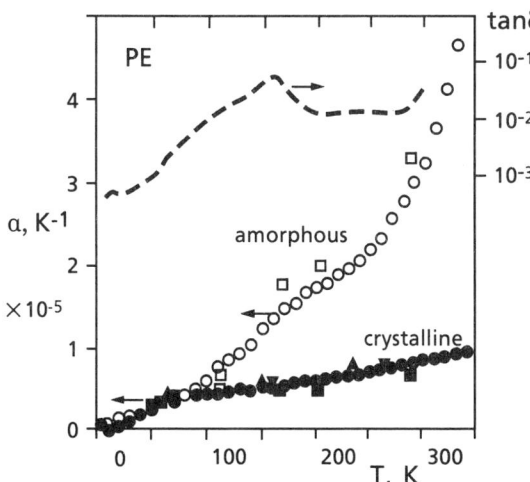

Fig. 4.14. Thermal expansion of amorphous and crystalline PE (extrapolated values) [4.5]. For comparison the damping spectrum $\tan\delta$ of semicrystalline PE at 10Hz is added.

Very different is the situation for PE extrapolated to its fully amorphous phase. There exists a strong glass transition, which gives rise to a large increase of a. The large thermal expansion of PE at higher temperatures has its origin in the influence of glass transitions. The thermal expansion curves of this section are combined with low-frequency damping spectra of the loss-factor $\tan\delta$, whose peaks roughly mark the glass transitions. Qualitatively it can be seen that a correlation exists between a and $\tan\delta$. (See also Appendix 4A). Most semicrystalline polymers exhibit a strong glass transition in the amorphous phase. Measurements have been performed on a series of polymers which show successive higher glass transition temperatures T_g. In Figure 4.15 the thermal expansion of PE, PTFE, POM and PP is plotted. It is clearly seen that a drastic rise of a is correlated with T_g. In this series, PP has the highest T_g, and hence the basic thermal expansion is maintained over a rather large temperature range. PE, on the contrary, has a lower T_g, and the contribution from the glass transition coincides with the beginning of the temperature independent region of a. The behaviors of PTFE and POM are intermediate.

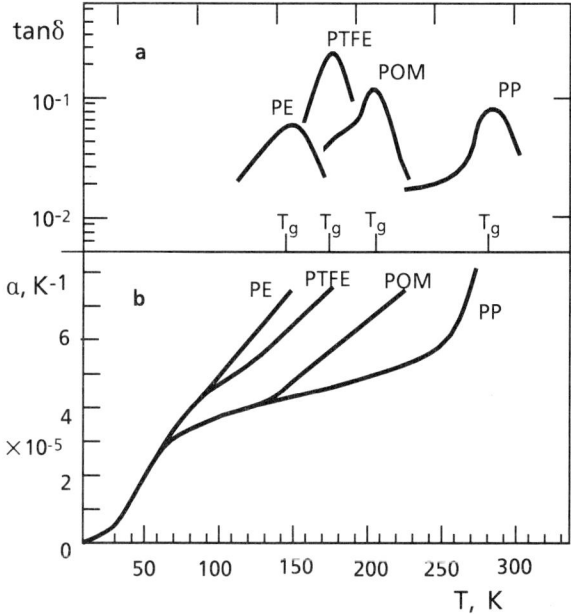

Fig. 4.15. Thermal expansion a of semicrystalline polymers with different glass transition temperatures T_g. a) Loss-factor $\tan\delta$ versus T. b) Coefficient of thermal expansion a versus T.

For amorphous polymers the influence of secondary and especially of tertiary glass transitions is small at high temperatures. Their contributions, however, become visible if they occur at low temperatures, where a generally is smaller. One example already has been given on PS. Other examples are shown in Figure 4.16 with a series of polyacrylates: PMMA, PEMA and PBMA. As seen in Figure 4.16a, PEMA has a glass transition around 40K. Its value of a is larger than that of PMMA which has no tertiary transition T_{gt} in this range. PBMA has a transition at a higher T_{gt}, and a consequently increases at a higher temperature. Another example is given in Fig.4.16b for PEI compared to PAI. Only the first polymer has a pronounced glass transition at $\approx 40K$, and the thermal expansion is higher. It becomes clear that there is a correlation between T_g and a. The correlation is of qualitative nature since the rise of a is too smeared out versus temperature. More examples of the influence of glass transitions can be seen in Figure 4.24 and in Appendix 4A. It is difficult to calculate analytically the thermal expansion of polymers at a glass transition. Some theoretical information can, however, be obtained from the Grueneisen relation (see Section 4.6).

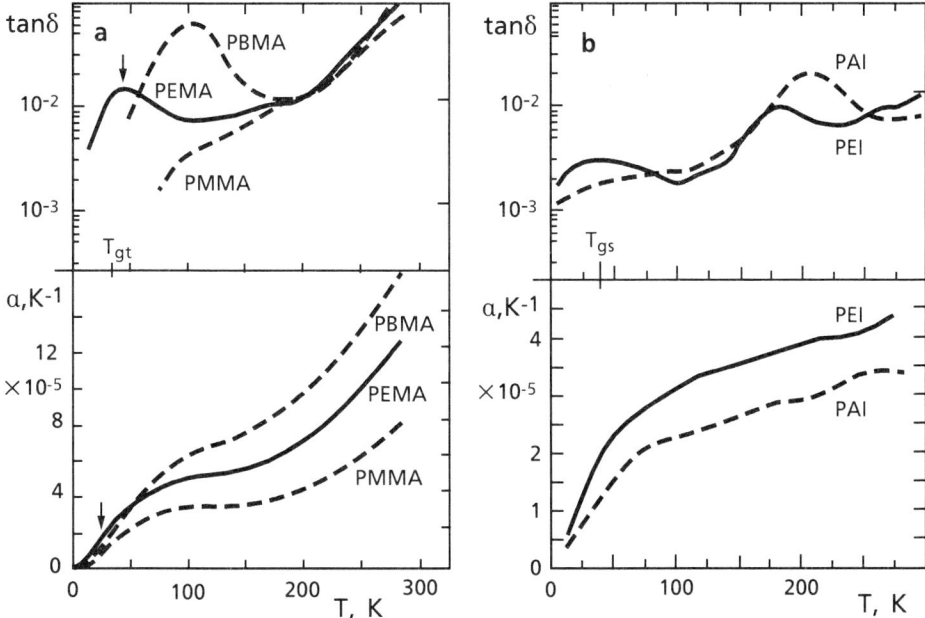

Fig. 4.16. a) Loss-factors tanδ versus T and coefficient of thermal expansion of several polyacrylates. b) Loss-factor and coefficient of thermal expansion of PAI and PEI. T_{gt} or T_{gs} are low temperature glass transitions.

4.4 THERMAL EXPANSION AND CROSS-LINKING

Density and functionality of cross-linking influence the thermal expansion of duroplastic polymers. Generally, the tendency is observed that thermal expansion decreases with decreasing cross-link distance. This is shown in Figure 4.17 for epoxy resins with different mean cross-link distances $\bar{d}$. The density of cross-linking has two main effects:

- Change of the relative contributions of vibrational modes and
- Shift of the glass transition temperatures.

A small cross-link distance suppresses interchain vibrations which are a source of a large thermal expansion. At the same time, the glass transition is shifted to higher temperatures. Both effects reduce the thermal expansion. In Figure 4.17 the integral thermal expansion $\Delta L/\Delta$ is plotted which reflects both effects. The reference temperature is RT. For resins (1) and (2) the glass transition is above RT, and does not influence the integral thermal expansion below RT. For resin (4), by contrast, the glass transition is near 300K, and a large contribution to $\Delta L/L$ results. The resins are cross-linked by three-functional groups, i.e., three chain segments are bound at a cross-link point. The range of $\Delta L/L$ from RT to 4.2K is from -0.9% to about -1.5%. Resins with four-functional groups of cross-linking show a lower thermal expansion.

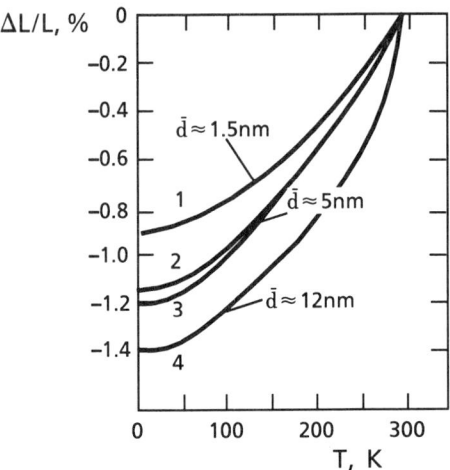

Fig. 4.17. Integral thermal expansion of epoxy resins with different cross-link distance $\bar{d}$.

4.5 GRUENEISEN RELATION

The Grueneisen relation correlates thermal expansion a with specific heat c and elasticity (bulkmodulus K) of a solid. The modulus K changes at glass transitions in a well-known manner. This can be used for calculating a at a glass transition from Eq.(4.3).

4.5.1 Theoretical Background of Single-mode Grueneisen Relation

In a first step, it is assumed that only one mode (index n) dominates the properties involved. Thermal expansion is given by the general expression [4.12]

$$3a_n \equiv \beta_n = \frac{dV}{dT}\frac{1}{V} = -\frac{1}{V}\left(\frac{\partial V}{\partial p}\right)_T \left(\frac{\partial p}{\partial T}\right)_V = K_n^{-1}\left(\frac{\partial p}{\partial T}\right)_V \qquad (4.15)$$

where K_n is the isothermal bulk modulus defined by

$$\frac{1}{V}\frac{\partial V}{\partial p} = -K_n^{-1} \qquad (4.16)$$

p is the thermal pressure, whose temperature dependence can be derived from the volume- (or pressure-) dependence of the specific heat [4.13, 4.14]. The specific heat can be expressed as a function of the frequency. It is postulated that the relative frequency shift per relative volume- (or pressure-) change is constant at all frequencies. The constant is the microscopic Grueneisen parameter defined in Eq.(4.4). The following relations are assumed

$$\gamma_n \equiv -\frac{d\omega_n}{\omega_n}\bigg/\frac{dV}{V} = \frac{dc_n}{c_n}\bigg/\frac{dV}{V} \qquad (4.17)$$

After a lengthy calculation one gets [4.8]

$$a_n = \frac{1}{3}\frac{c_n\,\rho}{K_n}\gamma_n \qquad (4.18)$$

The inverse specific volume V has been substituted by the density ρ. Specific heat depends on the harmonic components of oscillations while a results from its anharmonic components. The Grueneisen parameter γ_n is thus a measure of the anharmonicity of the oscillations (or their binding potentials).

The incorporation of the elasticity behavior (bulk modulus K) opens additional correlations:

Influence of Glass Transitions. Since c and the density ρ are not strongly affected by secondary or tertiary glass transitions it holds: $a_n \propto 1/K_n$. At a glass transition K drops and a_n rises. Experimental results show that the Grueneisen parameter is nearly temperature independent in the region of glass transitions. That means $K_n \cdot a_n \approx$ const. $= 1/3\, c_n \rho \gamma_n$. The constant can be measured, as well the behavior of K.

Pressure Dependence of Properties. The Grueneisen parameter reflects the volume- or pressure dependence of thermal vibrations. It can be used to evaluate the pressure dependence of several thermal and mechanical properties by means of the Grueneisen parameter [4.15 to 4.18].

- Sound velocity v: $\quad \gamma = K \dfrac{d \ln v}{dp} + \dfrac{1}{3}$ (4.19)

- Bulk modulus K: $\quad \gamma = 0.5 \dfrac{dK}{dp} - \dfrac{5}{12}$ (4.20)

- Thermal conductivity κ: $\quad \gamma = K \dfrac{d \ln \kappa}{dp} - \dfrac{1}{3}$ (4.21)

The reverse consideration is of more significance. If γ is known, one can estimate the pressure dependence of several properties, thus avoiding complicated investigations. The relative volume change can always be correlated to the pressure dependence by means of the bulk modulus K (see Eq.(4.16)).

4.5.2 Multi-mode Grueneisen Relation

In a general case several modes are involved. The partial Grueneisen parameters are weighted by the partial specific heat of each mode. The total mean Grueneisen parameter is defined by

$$\gamma_{tot} = \frac{\sum c_n \gamma_n}{\sum c_n} = \frac{\sum c_n \gamma_n}{c} \quad (4.22)$$

This yields the multi-mode Grueneisen relation

$$a = \frac{c\,\rho}{3K} \gamma_{tot} \quad (4.23)$$

The general Grueneisen relation incorporates all modes which determine a and c. If K and γ_{tot} are constant, the relation $a \propto c$ holds, and from multi-mode specific heat one can calculate the multi-mode a if γ_{tot} is known. Especially in more complicated cases it is easier to calculate specific heat instead of a (e.g., when intrinsic bending stiffness influences the thermal expansion).

The situation, however, is more complicated since generally γ_{tot} is a function of T. The partial Grueneisen parameters γ_n are assumed to be independent of temperature since only one mode determines a_n and c_n. This is not necessarily true for γ_{tot}. Each vibrational mode will have a different value of γ_n and will be dominant in a different temperature range. Thus, γ_{tot} becomes temperature dependent, even if all γ_n are constant. At low temperatures, interchain vibrations in the very asymmetric Van der Waals potential dominate. The correlated interchain Grueneisen parameter $\gamma_n = \gamma_{inter}$ will be large. At higher temperatures intrachain vibrations dominate with a small $\gamma_n = \gamma_{intra}$ due to the less asymmetric covalent potential. The determination of γ will be treated in Section 4.7. Some results on isotropic polymers are shown in Figure 4.18.

4.5.3 Grueneisen Tensor of Anisotropic Polymers

For anisotropic polymers not the bulk modulus K, but the elements of the elasticity tensor E_{ij} are used for describing the Grueneisen tensor [4.7; 4.8].

$$\{\gamma\} = E_{ij}\{a\}/(c\,\rho) \tag{4.24a}$$

The reversed relation is gained from the compliance tensor C_{ij}.

$$\{a\} = C_{ij}\{\gamma\}\,c\,\rho \tag{4.24b}$$

For an orthorhombic system with equal parameters in both directions perpendicular to the alignment [4.8]:

$$\gamma_\parallel = \left\{ 2E_{13}\,a_\perp + E_{33}\,a_\parallel \right\}/(c\,\rho) \tag{4.25a}$$

$$\gamma_\perp = \left\{ (E_{11} + E_{12})\,a_\perp + E_{13}\,a_\parallel \right\}/(c\,\rho) \tag{4.25b}$$

$$a_\parallel = \left\{ 2C_{13}\,\gamma_\perp + C_{33}\,\gamma_\parallel \right\} c\,\rho \tag{4.26a}$$

$$a_\perp = \left\{ (J_{11} + C_{12})\,\gamma_\perp + C_{13}\,\gamma_\parallel \right\} c\,\rho \tag{4.26b}$$

It should be mentioned that $\gamma_\perp$ in most cases is positive even if $a_\parallel$ in Eq.(4.26b) has a negative value [4.19].

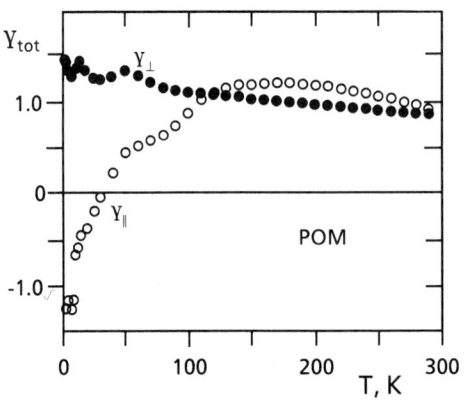

Fig. 4.17. Anisotropic components of the Grueneisen parameter for monocrystalline POM.

4.5.4 Measurement and Results of the Grueneisen Parameter

Determination from Thermal Data. The Grueneisen parameter can be determined from c, a and K using Eq.(4.23). Results of some amorphous and semi-crystalline polymers are plotted in Figures 4.18a and 4.18b. For comparison, theoretical results are given in Figure 4.18c. The experimental results, especially those of the amorphous polymers, are unusual. They show small values of γ at very low temperatures where interchain vibrations are expected to yield a large γ. The reason is that amorphous polymers have a very small γ_{tot} at very low temperatures owing to tunneling vibrations, which increase specific heat below 1K but not a. (see Fig.4.20). Maxima of γ_{tot} exist in between the regions where tunneling processes and intrachain vibrations are dominant.

At elevated temperatures a small γ_{tot} is expected for amorphous and crystalline polymers because of intrachain contributions. The strong decrease starting already at about 15K, however, is unexpected since intrachain vibrations only become dominant above about 70K. This behavior results from the bending (transverse) vibrations, which start contributing above 15K (see Fig. 3.4b). They increase specific heat much more than the thermal expansion; (bending vibrations even cause negative components of thermal expansion).

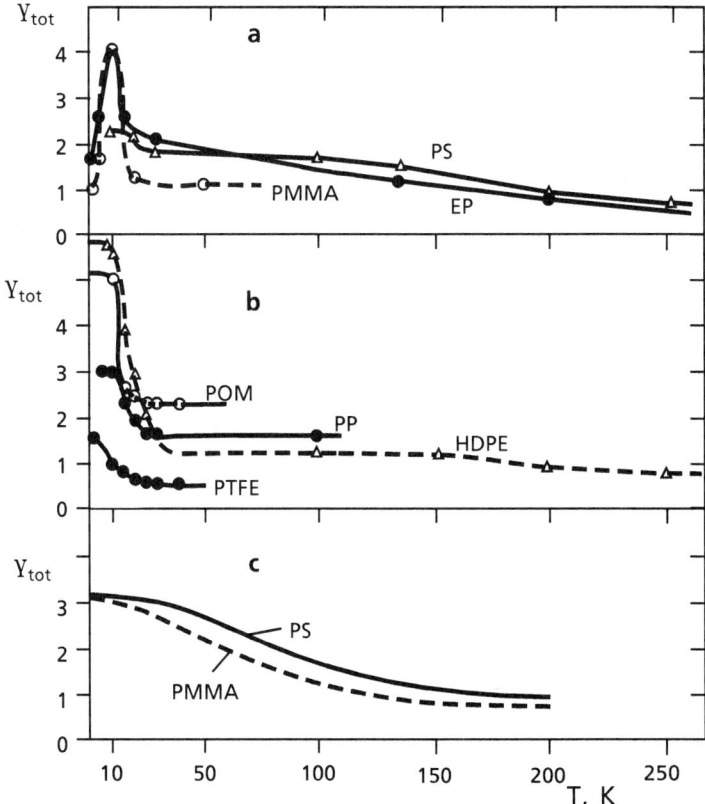

Fig. 4.18. Experimental results of the Grueneisen parameter γ_{tot}. **a)** amorphous polymers; **b)** semicrystalline polymers; **c)** theoretical curves [4.4].

The theoretical curves from Figure 4.18c show different values since only stretching vibrations are considered. The calculations are based on the relative frequency shift of stretching vibrations in a suitable potential with a relative volume- (or pressure-) change [4.4]. They again show larger interchain values of γ at low temperatures and smaller intrachain values at larger temperatures, but the temperature dependence is different and not consistent with experimental results. For an exact description the contributions from bending vibrations have to be incorporated. Above 40K, the values of γ are rather independent of temperature despite the fact that secondary or tertiary glass transitions can cause an increase of a. The Grueneisen parameter is not influenced by glass transitions. At elevated temperatures it behaves similarly for semicrystalline and amorphous polymers since intrachain vibrations are rather independent of the crystalline fraction.

Ultrasonic Attenuation. The Grueneisen parameter can also be obtained from the frequency dependence of ultrasonic attenuation [4.19; 4.20]. This method yields experimental values, which cover a large range for polymers:

$T \approx 20K : \gamma \approx 8$ to 13
$T = 77K : \gamma \approx 2$ to 12
$T = 240K : \gamma \approx 0.6$ to 5

Again, the Grueneisen parameter is much larger at low temperatures.

Thermoelastic Effect. This effect is the thermodynamic inversion of thermal expansion in that external pressure applied on a solid results in a temperature change [4.21; 4.22]. It can be shown that [4.23]

$$\gamma = -\frac{3E}{T}\frac{dT}{d\sigma} \qquad (4.27)$$

σ is a uniaxial tension applied under adiabatic conditions and E is the modulus. The γ values found by this method are at least consistent with those from thermal data.

4.6 TEMPERATURE DEPENDENCE OF THE THERMAL EXPANSION AND OF THE GRUENEISEN PARAMETER

For both the thermal expansion and the Grueneisen parameters significant differences exist between amorphous and crystalline polymers. The temperature dependence is greatly influenced by the crystalline content. These properties will be treated in the characteristic temperature ranges where specific processes prevail.

Temperature Range: T < 4K (Tunneling Range)

Tunneling vibrations in amorphous polymers increase specific heat more than thermal expansion and consequently the Grueneisen parameter decreases at very low temperatures. An example is given for PMMA in Figure 4.19.

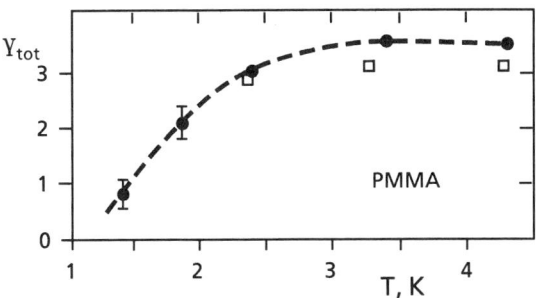

Fig. 4.19. Grueneisen parameter for PMMA [4.21].

For highly crystalline polymers, by contrast, no tunneling processes occur and the relation $a \propto T^3$ is valid even at very low temperatures. This is shown in Figure 4.20 for PTFE with 85% crystallinity. Nearly the same temperature dependence has been found for the specific heat. Since K is temperature independent, the Grueneisen parameter is nearly constant for highly crystalline polymers and $\gamma \propto a/c \approx$ const [4.24; 4.25].

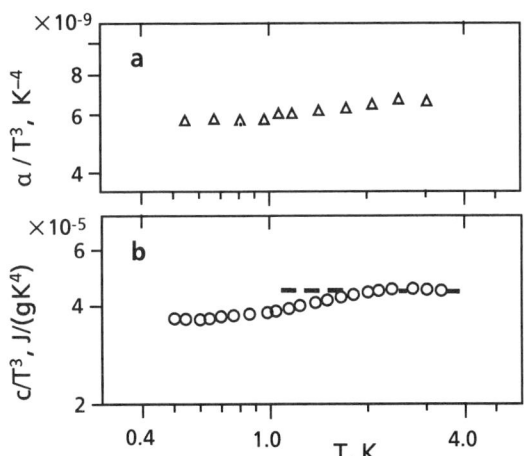

Fig. 4.20. Thermal expansion and specific heat of PTFE at low temperatures [4.15; 4.26]. a) ratio a/T^3; b) ratio c/T^3.

Temperature Range: T ≈ 5K to 30K (Debye Range)

Above 5K contributions from tunneling effects are negligible. Up to 30K the Debye relation $a \propto T^3$ holds for fully crystalline polymers. Less crystalline polymers show a different behavior. This is demonstrated in Figure 4.21. The ratio a/T^3 is plotted versus temperature for HDPE with different crystalline fractions f_c. Highly crystalline HDPE ($f_c \to 1$) shows the relation $a \propto T^3$. Since an analogous relation has been found for the specific heat and K is nearly constant, the Grueneisen parameter should be independent of temperature in this temperature range.

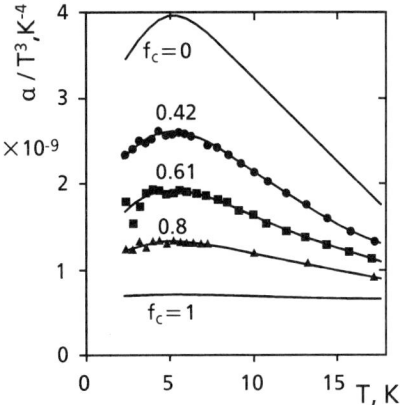

Fig. 4.21. Ratio a/T^3 versus temperature for HDPE with different crystalline fractions f_c. The curves with $f_c=0$ and $f_c=1$ are extrapolations [4.5; 4.6].

The amorphous phase ($f_c=0$) shows a different behavior. The ratio a/T^3 depends on temperature and goes through a maximum in the vicinity of 5K. This values of a depend on temperature stronger than T^3 below this maximum and weaker above. This behavior enhances the maximum of γ_{tot} found for amorphous polymers. Above 10K, the temperature dependence of a becomes weaker for amorphous polymers and to a lesser degree for semicrystalline ones. In the range 10K to 50K it holds for

semicrystalline polymers: $a \propto T^{1.7}$ to T^2
amorphous polymers: $a \propto T^{1.3}$ to $T^{1.7}$

In Figures 4.22a to 4.22c the temperature dependences of a for several amorphous and semicrystalline polymers (crystalline content about 60 vol.%) are plotted [4.27; 4.28]. It appears that the thermal expansion coefficient of the semicrystalline polymers depends less on the chemical compositions than that of

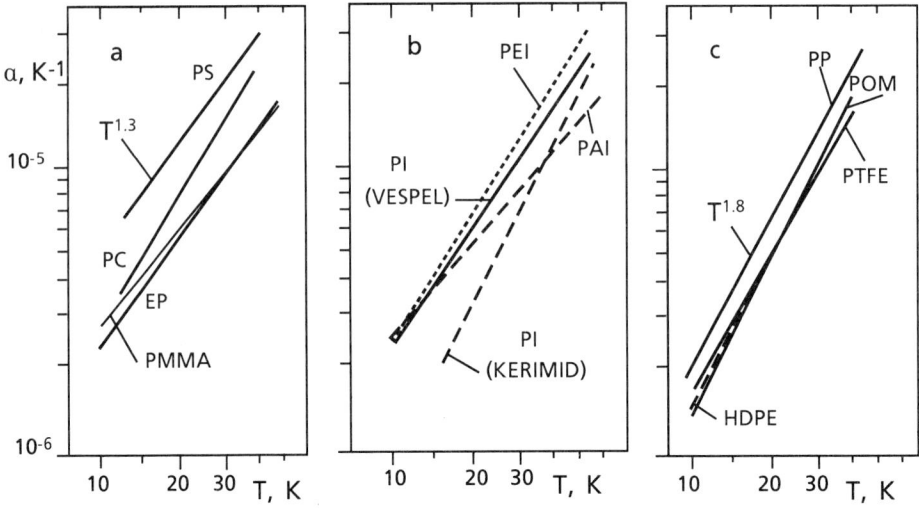

Fig. 4.22. Thermal expansion a versus temperature. **a)** and **b)** amorphous, **c)** semicrystalline.

the amorphous ones. The temperature dependence is also more similar for semi-crystalline polymers.

Temperature Range: 30K to 80K

In Figure 4.23 the thermal expansion for several polymers is compiled. (More data are provided in the Appendix 4 A). Below 70K, the thermal expansion is a rather strong function of temperature. The semicrystalline polymers show a similar thermal expansion independent of their chemical compositions.

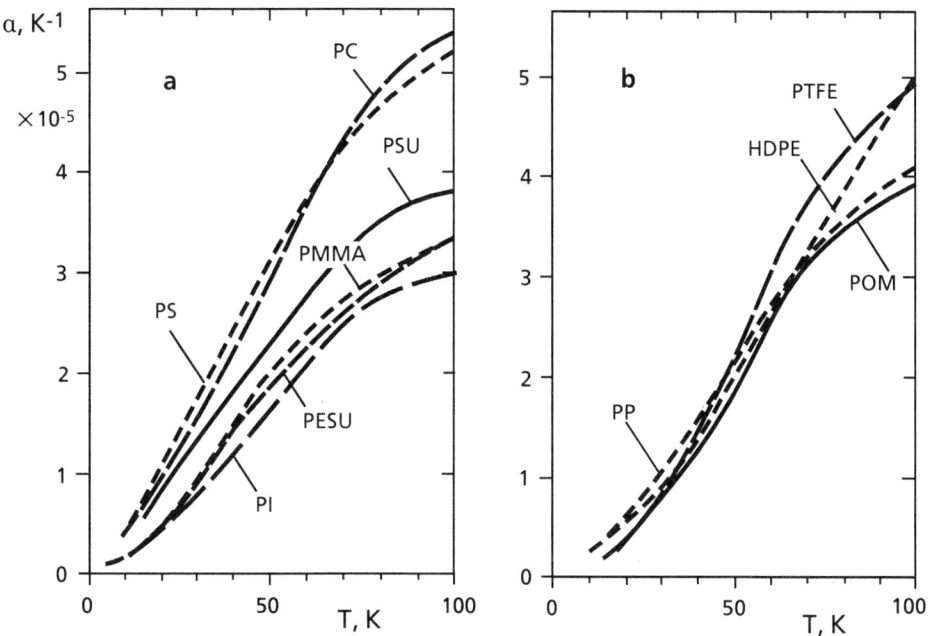

Fig. 4.23. Coefficient of thermal expansion a versus temperature.
a) amorphous; b) semicrystalline.

Temperature Range: 80K to 300K (Glass Transition Range)

Above about 130K the influence of glass transitions becomes dominant. The thermal expansion of several amorphous polymers is shown in Figures 4.24a to 4.24d together with damping spectra. The correlation between glass transitions and thermal expansion is evident. The Grueneisen parameter, however, is similar for amorphous and semicrystalline polymers in this temperature range (see Fig. 7.18). This is because the Grueneisen parameter, as defined by Eq.(4.3), is not sensitive to secondary or tertiary glass transitions. In the temperature range 80K to 300K the Grueneisen parameter is not very temperature-dependent.

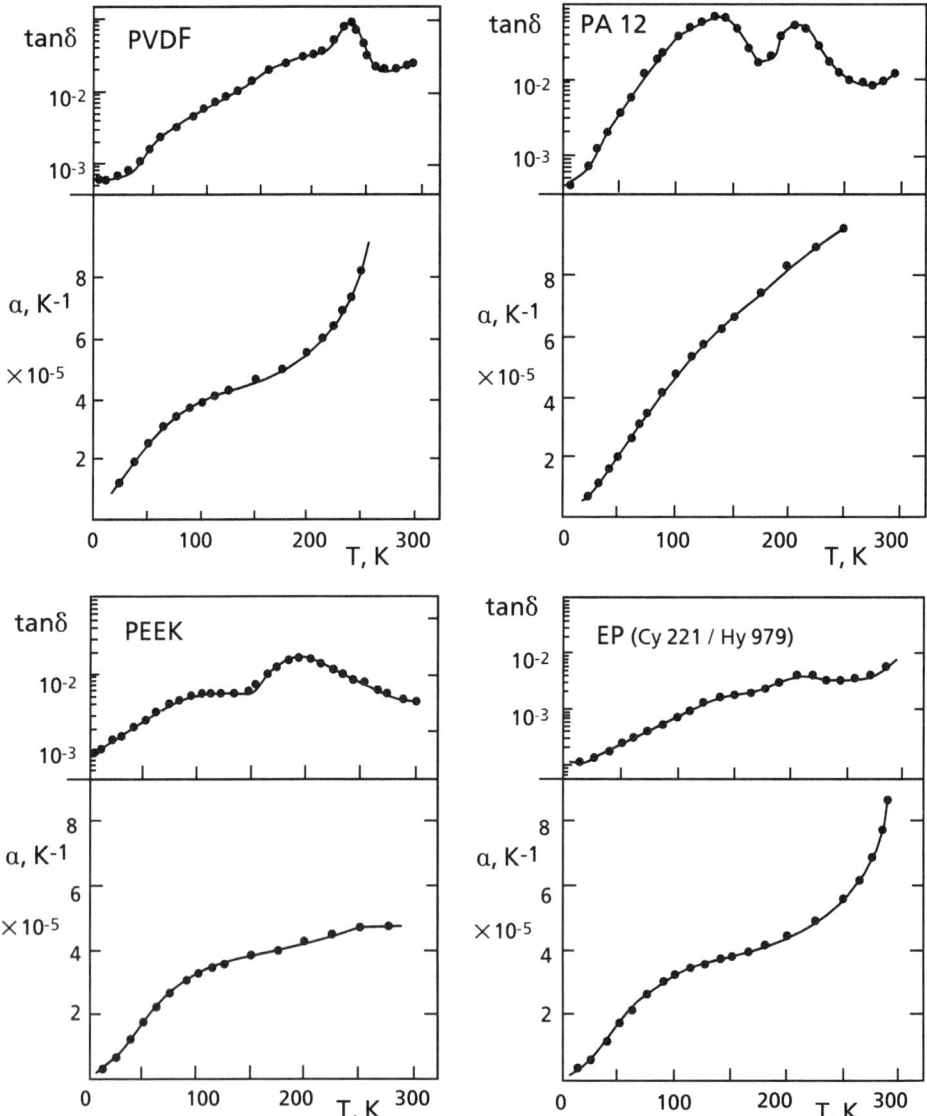

Fig. 4.24. Coefficient of thermal expansion versus temperature for several polymers. The damping spectra are added for comparison. More data are compiled in Appendix 4A.

4.7 SUMMARY

(1) Thermal expansion of polymers originates from different modes of vibrations (basic thermal expansion) and is strongly influenced by glass transitions, in amorphous polymers.

(2) Longitudinal (stretching) vibrations in asymmetric potentials lead to a positive thermal expansion:
- Intrachain vibrations in the covalent potential yield a small $a_{\parallel s}$
- Interchain vibrations in the Van der Waals potential yield a large $a_{\perp sw}$.

(3) Transverse (bending) vibrations lead to a small, but negative thermal expansion $a_{\parallel b}$ in the chain direction.
Perpendicular to the chain a large positive $a_{\perp b}$ results which, however, is somewhat different from $a_{\perp sw}$ where the bending stiffness of chains is included.

(4) The simple components are combined into a total thermal expansion:

$$a_{\parallel tot} = a_{\parallel} \approx a_{\parallel s} + a_{\parallel b}$$
$$a_{\perp tot} = a_{\perp} \approx a_{\perp sw} + a_{\perp b}$$

The anisotropic expansion behavior can be studied on aligned or monocrystalline polymers and on self-orienting LCP.

(5) Isotropic polymers are entangled and a mean thermal expansion can be calculated from the above contributions:

$$a_{iso} \approx \tfrac{1}{3}(a_{\parallel} + 2\,a_{\perp}).$$

(6) The thermal expansion arising from vibrational modes forms a "basic thermal expansion". It shows a strong temperature dependence up to ~70K from interchain contributions. Above this temperature, intrachain vibrations prevail which show only a weak temperature dependence.

(7) Thermal expansion of amorphous polymers is increased by glass transitions; there is a superposition of the basic thermal expansion and the contributions from glass transitions. For tertiary or secondary glass transitions a contribution is visible only in the low temperature range where a is small. At higher temperatures, only strong glass transitions yield a noticable contribution. A pronounced effect can be seen on semicrystalline polymers with a strong glass transition in their amorphous phase. The temperature dependence of a is therefore different for amorphous and highly crystalline polymers.

(8) Thermal expansion of duroplastic polymers is influenced by the density and functionally of cross-linking. Small cross-link distances suppress interchain vibrations which are known to contribute to a large coefficient of thermal expansion. At the same time the glass transitions are shifted to higher temperatures. Both effects decrease the thermal expansion. Duroplastic polymers with four-fold functionality cross-linkage exhibit a lower thermal expansion than three-fold ones.

(9) A correlation between the thermal expansion a, the specific heat c and the bulk modulus K is given by the Grueneisen relation.

$$a = \frac{1}{3}\frac{\rho}{K} c\,\gamma$$

If only one mode of vibration controls a, c and K, then γ is constant. For polymers several modes are involved, and a total Grueneisen parameter γ_{tot} is defined which in general is a function of T.

(10) The Grueneisen parameter is a measure of the asymmetry of the binding potential involved or the anharmonicity of oscillations. For harmonic vibrations in a symmetric binding potential, no thermal expansion occurs.
At low temperatures vibrations in the rather asymmetric Van der Waals potential determine the thermal expansion and hence γ_{tot} is large. For amorphous polymers below 1K, however, the Grueneisen parameter is drastically decreased because tunneling vibrations increase the specific heat but not the thermal expansion. At higher temperatures, intrachain vibrations in the less asymmetric covalent potential prevail and γ_{tot} becomes small. In addition to that, bending vibrations contribute which yield a lower Grueneisen parameter.

(11) It is worth mentioning that the Grueneisen parameter is hardly influenced by glass transitions. The reason is a compensation of a and K, while c is not much influenced. At a glass transition a gets larger and K smaller (dispersion step). Thus, the Grueneisen parameter stays nearly constant in the dispersion region.

(12) The Grueneisen relation of aligned polymers is described by a Grueneisen tensor.

(13) The pressure dependence of several mechanical and thermal properties can be estimated by the Grueneisen parameter.

Summary of Contributions to Thermal Expansion.

		Sign	Value
1) **Basic Expansion**			
Longitudinal vibrations	Intrachain	positive	low
	Interchain	positive	high
Transverse vibrations	$\parallel$	negative	low
	$\perp$	positive	high
Combined	$a_\parallel$		
	$a_\perp$		
For isotropic, entangled polymer chains $a \ll 0$			
2) **Expansion at Glass Transitions**			
Change of free volume / Basic expansion		positive	high

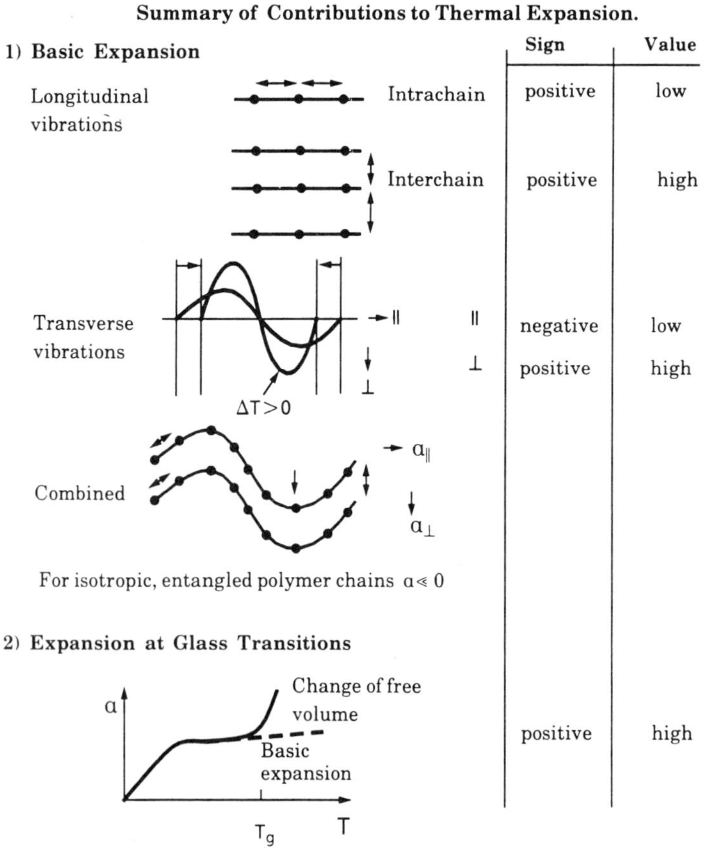

Fig. 4.25. Thermal expansion of polymers

4.8 REFERENCES

4.1 Kittel, Ch; Introduction to Solid State Physics; p.228; John Wiley and Sons, New York (1973).
4.2 Barron, T,.H.K., Collins, J.G. and Shite, G.K.; Adv. in Phys. **29** (1980); p.609.
4.3 Wada, Y.A., Itani, T. Nishi and Nagai, S.; Polym. Sci. **7** (1969), p.201 (Part A2).
4.4 Curro, J.G.; J. Chem Phys. **58** (1973); p.374.
4.5 Engeln, J. and Meissner, M. in "Nonmetallic Materials and Composites at Low Temperatures ", Vol. 2; p. 1, Eds.: Hartwig, G. and Evans, D., Plenum Press (1982).
4.6 Engeln, J.; Thesis, TU Berlin, (1983), D 83, p.243.
4.7 White, G.K. in: "Nonmetallic Materials and Composites at Low Temperatures ", Vol. 3; p.1, Eds.: Hartwig, G. and Evans, D., Plenum Press (1986).
4.8 White, G.K. and Choy, C.L.; J. Polym Sci. **22** (1984); p. 843 (Phys. Ed.).
4.9 Smith, K. J. Jr.;Polymer Engineering and Sci. **24** (1984); p.205.
4.10 Choy, C.L., Chen, F.C. and Ong, E.L.; Polymer Sci. **20** (1979); p.1191.
4.11 Simha, S.C. and Boyer, R.F.; J. Chem. Phys. **37** (1962); p.1003.
4.12 Ashcroft, W.N. and Mermin, N.D.; "Solid State Physics" Cornell University.
4.13 Busch, G. and Schade, H.; Vorlesungen über Festkörperphysik; Birkhauser Verlag, Basel (1973).
4.14 Barker, R.E. Jr. and Chen, R.Y.S.; J. Chem Phys. **53** (1970); p.2616.
4.15 Matsushige, K., Hirakawa, S. and Takemura, T.; Memoirs of the Faculty of Engineering, Kyushu Uni., **Vol. 32** (1972); p.153.
4.16 Sharma, B.K.; Polymer **24** (1983);p.314.
4.17 Wu, C.K., Jura, G. and Shen, M.; J. Appl. Phys. **Vol. 43** (1972); p.4348.
4.18 Chen, F.C., Choy, C.L., Wong, S.P. and Young, K.; J. Polymer Sci. **19** (1981); p.971.
4.19 Yamamoto, K. and Wada, J. Y.; Phys. Soc. Jpn. **11** (1956); p.887.
4.20 Federle, G. and Hunklinger, S. in "Nonmetallic Materials and Composites at Low Temperatures ",Vol. 2; p.49, Eds.: Hartwig, G. and Evans, D., Plenum Press (1982).
4.21 Wright, O.B. and Phillips, W.A.; Philosoph Mag. **B 50** (1984); p.63.
4.22 Tietje, H., Schickfus, .v.M and Gmelin, E.; Z. Phys. B - Condensed Matter, **64** (1986); p.95.
4.23 Schickfus, M.v., Hunklinger, S. and Transfeld, K. in "Nonmetallic Materials and Composites at Low Temperatures ", Vol. 2; p. 37, Eds.: Hartwig, G. and Evans, D., Plenum Press (1982).
4.24 Wang, Li-Hui, Choy, C.L. and Porter, R.S.; J. Polymer Sci. **20** (1982); p.633.
4.25 Wang, Li-Hui, Choy, C.L. and Porter, R.S.; J. Polymer Sci. **21** (1983); p.657 (Polymer Phys. Ed.).
4.26 Boyer, J.D., Lasjaunias, J.C., Fisher, R.A. and Phillips, N.E.; J. Non Crist. Solids **55** (1983); p.413
4.27 Schwarz, G.; Cryogenics **28** (1988); p.248.
4.28 Hartwig, G.; Cryogenics **28** (1988); p.255.

THERMAL CONDUCTIVITY

Contents

5.	Thermal conductivity	99
5.1	Definition and survey	99
5.2	General theory	100
5.3	Theory for amorphous polymers	102
5.4	Theory for semicrystalline polymers	105
5.5	Results and discussion	110
5.6	General features of thermal conductivity of polymers	111
5.7	Thermal diffusivity	114
5.8	Summary	115
5.9	References	116
Appendix		
5A	Data	

5. THERMAL CONDUCTIVITY

5.1 DEFINITION AND SURVEY

Thermal conduction comprises essentially that portion of phonon transport in which, different from sound propagation, the frequency, the phase and the polarization of the phonons are not maintained. Thermal conductivity is a process in which an inhomogeneous thermal excitation initiates various thermodynamic relaxation processes and thus gives rise to a more or less slow transport of energy with permanent local thermalization. The resulting temperature gradient drives the flux of energy carriers which in the case of insulators are phonons. The relationship between the flux of heat power $\dot{Q}$ per area A and the temperature gradient[1] is given by the coefficient of thermal conductivity κ.

$$\dot{Q}/A = -\kappa \cdot \text{grad } T \; ; \; W/m^2 \tag{5.1}$$

In case of anisotropic materials (stretched polymers, fibers) the relation is given by the conductivity tensor |κ| [5.1; 5.2; 5.3]. In the nonsteady case, both $\dot{Q}(t)$ and grad T(t) are functions of time. Usually, κ is defined in the steady state in which a thermal equilibrium is established at each location.

Thermal conductivity means that the heat flux is not transported via direct transfer mechanisms such as convection or radiation but via vibrations (i.e., phonons) which are thermalized by various scattering processes. Thermalization means the establishment of a local thermal equilibrium. The amount and type of scattering processes determines the thermal resistance and its temperature dependence.

For polymers several thermal scattering processes are working in different temperature ranges:

T≤ 1K: resonant scattering by tunneling processes in amorphous polymers;
T≥ 1K: defect- or structure scattering; scattering at local density variations or scattering at cross-link points of thermosets;
T≤20K: interface scattering at crystallites of semicrystalline polymers.

---

[1] The gradient is formally of negative sign since the temperature and the length are counted in opposite directions.

A schematic survey of thermal conductivity as a function of temperature is given in Figure 5.1.

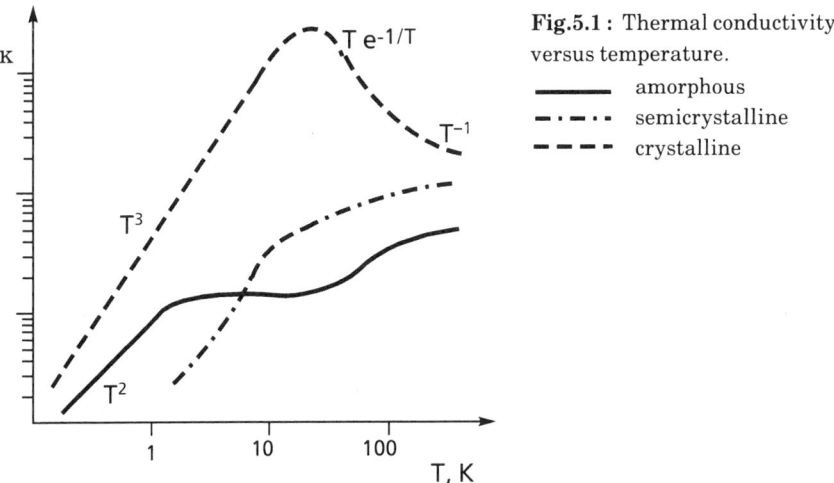

Fig.5.1 : Thermal conductivity versus temperature.
——— amorphous
—·—·— semicrystalline
— — — crystalline

The thermal conductivity of monocrystals obeys a T^3 dependence at low temperatures; then it decreases after having reached a maximum due to additional thermal resistivities from Umklapp processes ($\kappa \propto T e^{-1/T}$) and phonon-phonon scattering ($\kappa \propto 1/T$). Polymers, in most cases, are not crystals and do not show this behavior. Phonon-phonon scattering and Umklapp processes are negligible since they are dominated by the scattering mechanisms mentioned before. Amorphous polymers show a T^2 dependence of the thermal conductivity at very low temperatures which is caused by phonon scattering at tunneling systems. In the interval 5K to 15K a plateau is typical of amorphous polymers. Above 15K the temperature dependence is generally small. Semicrystalline polymers exhibit a larger thermal conductivity at high temperatures. At low temperatures, however, it drops below that of amorphous polymers because of interface scattering at crystallite surfaces.

5.2 GENERAL THEORY

Thermal conductivity depends on the vibrational energy which can be activated at a temperature T. This term is given by the specific heat c. The next important parameter for phonon propagation is the mean phonon velocity $\bar{v}$. It does not have a large temperature or frequency dependence, especially not at low temperatures. The mean phonon velocity $\bar{v}$ is given by the longitudinal and transverse phonon velocities (see Section 2.5).The most sensitive parameter of thermal conductivity is the phonon mean free path l. It is the path travelled without obstruction by a phonon between two scattering processes. Since many scattering events occur, the mean statistical value is used. The phonon path is inversely proportional to the thermal resistivity caused by a scattering mechanism. If the thermal resistivities are additive for different scattering mechanisms, the free phonon paths sum up reciprocally.

THERMAL CONDUCTIVITY

A relation, extrapolated to a "phonon gas," is taken as the basis for this calculation [5.1].

$$\kappa(T) = \frac{\rho}{3} \int_0^{\omega_{max}} c_\omega(\omega,T)\, l(\omega,T)\, v(\omega,T)\, d\omega\,;\ W/(m,K) \qquad (5.2)$$

The term c_ω indicates the contribution to specific heat per frequency interval from ω to $\omega + d\omega$. The mean velocity $\bar{v}$ is rather independent of frequency and therefore it holds:

$$\kappa(T) \approx \frac{\rho\bar{v}(T)}{3} \int_0^{\omega_{max}} c_\omega(\omega,T)\, l(\omega,T)\, d\omega\,;\ W/(m,K) \qquad (5.3)$$

The decisive point is the combined frequency integration of both, c_ω and l. Phonon scattering processes may cause a particular frequency dependence, which disturbs the usual frequency-temperature relation given by the probability of states (Section 2.6). The frequency- and temperature dependence is not always known for l. A well-known approximation neglects the frequency dependence.

$$\kappa(T) \approx \frac{\rho}{3} c(T)\, \bar{l}(T)\, \bar{v}(T)\,;\ W/(m,K) \qquad (5.4)$$

Frequency-averaged parameters are denoted by a dash. Equation (5.4) looks simple but the difficulties are hidden in the evaluation of the mean free path $\bar{l}$. Only in special cases a direct application of Eq. (5.4) is possible. One of those systems are monocrystals in which no scattering processes occur at very low temperatures except on the surfaces of the crystal. At a sufficient low temperature only long wavelength phonons are activated, whose mean free path is determined by the size of the crystal and thus, $l \approx$ const. At low temperatures the phonon velocity v is constant and the specific heat $c \propto T^3$. From Eq.(5.4) it follows

$$\kappa(T) \propto T^3 \quad \text{for}\ T < 10K \qquad (5.5)$$

For polymers, the situation is more complicated. The physical processes which govern thermal conductivity are very different for crystalline, semicrystalline and amorphous polymers. Therefore, they will be discussed separately for these classes of polymers. The temperature dependence of the thermal conductivity is very different within certain temperature ranges. A rough compilation is given in Table 5.1.

Table 5.1. Temperature dependence of thermal conductivity.

Temperature range	T dependence	Assumed scattering process
Amorphous Polymers		
$T \leq 1K$	$\kappa_a \propto T^2$	Scattering at tunneling systems.
$T \approx 4K - 10K$	$\kappa_a \propto$ const.	Rayleigh scattering.
$T > 30K$	$\kappa_a \propto T^m$ m : 0.3 – 0.5	Defect- and structure scattering.
Semicrystalline Polymers		
$T < 20K$	$\kappa_{semi} \propto T^2$	Interface scattering at crystallites.
$T > 30K$	κ_{semi} is determined mainly by $\kappa_a(T)$	

The next two sections will provide the theoretical background for understanding those dependencies.

5.3 THEORY FOR AMORPHOUS POLYMERS

Contrary to crystallites, the phonon mean free path in amorphous polymers at low temperatures generally depends on both, the frequency and the temperature. Therefore, the thermal conductivity of amorphous polymers exhibits a different temperature dependence, which will be considered in the following temperature ranges:

Temperature Range T ≤ 1K (Range of Tunneling Processes)

In this temperature range the main scattering processes are due to phonon absorption and reemission after a certain relaxation time by tunneling systems (two-level systems). The scattering mechanism considered is resonant tunneling absorption of those phonons which have an energy just equal to the level difference. The mean free path l for this process is a function of frequency and temperature [5.4]:

$$l \propto \omega^{-1} \coth(\hbar\omega/2 k_B T) \tag{5.6}$$

Amorphous polymers have a broad energy distribution of two-level tunneling systems, so that phonons at a broad frequency range are involved. An approximation is made which relates frequency and temperature. The main contributions are made by phonons with an energy $\hbar\omega \approx 2 k_B T$ (dominant phonons). From Eq.(5.6) it results

$$l \equiv l(\omega) \propto \omega^{-1} \quad or \quad l \equiv l(T) \propto T^{-1} \tag{5.7}$$

As already mentioned, tunneling systems determine the thermal conductivity in this range. In a first attempt the temperature dependence of κ is calculated by using the following relations:

- specific heat : $c \propto T$ (main contribution from tunneling systems)
- mean free path : $l \propto T^{-1}$
- phonon velocity : $v = $ const.

From Eq.(5.4) one gets $\kappa = $ const. This is in contradiction to experimental results which yield $\kappa(T) \propto T^2$ (see Figure 5.10a). Remember, a more accurate description of specific heat has been given in Chapter 3 by Eq.(3.13).

$$c = gT + hT^3 \tag{5.8}$$

A result, $\kappa \propto T^2$ is obtained when using only the small second term.[2] This is roughly in agreement with experiments. An interpretation of this strange phenomenon has been given by Zaitlin and Anderson [5.5]. They assume that tunneling vibrations are localized and do not contribute to phonon propagation. Tunneling systems, however, act as thermal resistances and determine the mean free path l. Tunneling vibrations store internal energy and increase the specific heat but do not increase the thermal conductivity. One more refinement is necessary for an exact description. As mentioned in Section 3.1, the specific heat of amorphous polymers is increased by localized vibrations which contribute at low temperatures Even the term hT^3 contains an extra term, which overestima-

---

[2] This term can be obtained by extrapolating the T^3 dependence below 1K.

tes thermal conductivity. The correct presentation of the specific heat is

$$c = g\, T_{loc} + (h_1\, T^3)_{loc} + (h_2\, T^3)_{Debye} \tag{5.9}$$

Only the nonlocalized Debye term contributes to thermal conductivity and yields a result which agrees with experiments [5.6].

$$\kappa(T) \propto T^2 \quad \text{for } T < 1K \tag{5.10}$$

Temperature Range: 5K to 15K (Plateau Region)

A peculiar feature of amorphous polymers is a plateau of the thermal conductivity occuring between 5K and 15K. Several explanations have been proposed. A clear physical explanation, however, is still missing. A formal solution can be provided by assuming that the mean free path $l(\omega)$ obeys a particular frequency dependence, which is chosen in such a way that from Eq (5.4) $\kappa(T) = $ const. It is hoped that this procedure will provide a hint about the particular scattering process involved. The differential specific heat c_ω of Eq.(5.3) is expressed by Eq.(3.2)

$$c_\omega \propto \frac{\delta}{\delta T}(\hbar\omega\, D(\omega) <m>) \tag{5.11}$$

The integral of Eq.(5.3) is then determined by the product $D(\omega)\cdot l(\omega)$ which can be expressed by an exponential law ω^m. The integral can be solved in analogy with Eq.(3.5) of the specific heat when substituting $D(\omega)$ by $D(\omega)\cdot l(\omega)$. For the acoustic approximation, $T \ll \Theta$, one gets a relation analogous to Eq.(3.8)

$$\kappa(T) \propto T^{m+1} \quad \text{if} \quad D(\omega)\, l(\omega) \propto \omega^m \tag{5.12}$$

An exponent $m = -1$ yields:

$$\kappa(T) \propto T^0 = \text{const.} \tag{5.13}$$

Since $D(\omega) \propto \omega^2$ it follows from Eq.(5.12)

$$l(\omega) \propto \omega^{-3} \tag{5.14}$$

The strong frequency dependence of the mean free path explains in a formal way the plateau region. A scattering process, however, with the required frequency dependence is not known. Rayleigh scattering is a process obeying a strong frequency dependence ($l \propto \omega^{-4}$). Anderson and Zaitlin have shown that even with this relation (instead of ω^{-3}) a reasonable plateau region could be explained [5.5]. In Figure 5.2 is shown schematically the frequency dependence of the mean free path l which is assumed for amorphous polymers. A sharp drop of l occurs from low to high frequencies. Amorphous polymers (and other glassy solids) seem to act as low pass filters for phonons in the plateau region. Quite novel considerations attribute the plateau to an energy gap between the phonon and fracton spectra in that range. Fractons are supposed to be quantized units of vibrations which are related to elementary fractal geometry [5.7; 5.8].

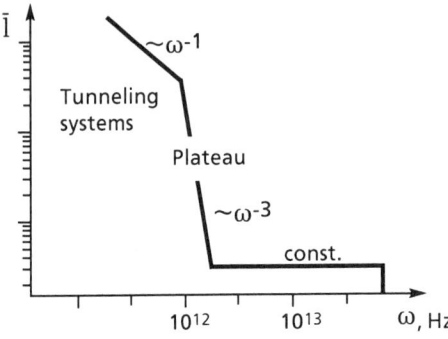

Fig. 5.2. Frequency dependence of the mean free path l of amorphous polymers (schematic).

Temperature Range: T > 30K

In this range a nearly frequency- or temperature-independent phonon path can be assumed. It is on the order of several atomic spacings.[3]

$$\bar{l} \approx \text{constant} \tag{5.15}$$

This assumption is consistent with the experimental results. In Figure 5.3, according to Eq.(5.4) the mean free path $\bar{l}$ is plotted as a function of temperature for epoxy resins, which differ by their mean cross-link distance $\bar{d}$. It can be recognized that $\bar{l}$ is hardly dependent on temperature above 30K. The values below 30K are only of qualitative nature, because they do not take account of any frequency dependence according to Eq.(5.3). They show qualitatively that the mean free path $\bar{l}$ depends on the cross-link distance $\bar{d}$ when the dominant phonon wavelength λ_D has the same order of magnitude. λ_D is representative of phonons which carry most of the heat at a temperature T (see Section 5.6).

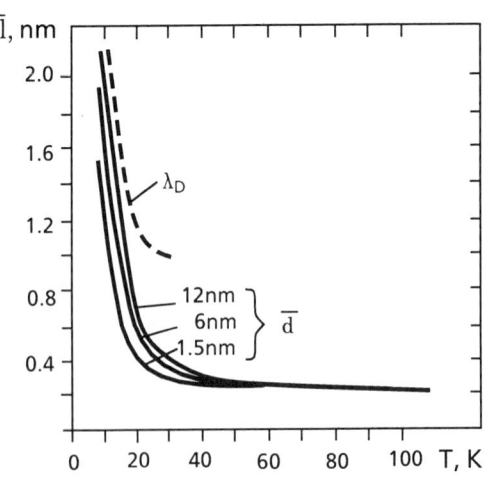

Fig. 5.3. Mean free path $\bar{l}$ for epoxy resins with different cross-link distances $\bar{d}$ versus temperature. The dashed curve represents roughly the dominant phonon wavelength λ_D.

Thermal conductivity of amorphous polymers is not very dependent on temperature above 30K. From experimental results one gets

$$\kappa \propto T^m \text{ with } m = 0.3 \text{ to } 0.5 \tag{5.16}$$

---

[3] The term phonon is hardly applicable when only some periods of vibrations occur within a mean free path.

5.4 THEORY FOR SEMICRYSTALLINE POLYMERS

Crystallites are embedded in amorphous domains. Their combined thermal resistances determine mean free phonon path of semicrystalline polymers; the main reduction of $\bar{l}$ occurs, of course, in the amorphous domains. A rough estimation of $\bar{l}$ is given in Figure 5.4 for PE with different crystalline fractions f_c. These values are not deduced from a physical process; they have been calculated by Eq.5.4 and serve only for illustrating the order of magnitude [5.9; 5.10].

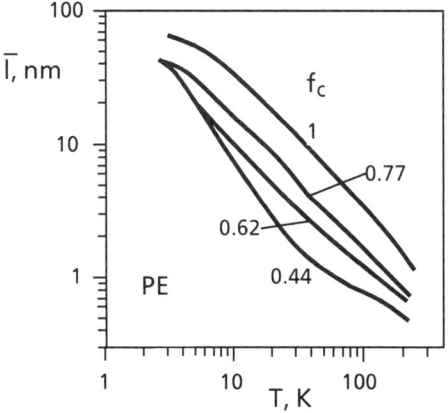

Fig. 5.4. Estimation of phonon mean free path $\bar{l}$ of semicystalline PE with different crystalline fractions f_c versus temperature [5.9]. (The values at the crystalline fraction of $f_c = 1$ are extrapolations).

The tendency can be seen that $\bar{l}$ is drastically reduced with decreasing f_c. The value of $\bar{l}$ becomes less dependent on temperature when f_c is decreased. For completely amorphous polymers a constant phonon mean free path can be assumed above 50K (see Figure 5.3).

5.4.1 Boundary Scattering (T<20K)

The crystallites exhibit a relatively high thermal conductivity, which can be considered as a thermal short cut in an amorphous environment. A thermal resistance by boundary scattering between phases becomes dominant below 20K. The resistance is reflected as a temperature step ΔT at the interface of the crystallites. A schematic diagram of the temperature profile of a semicrystalline polymer is shown in Figure 5.5. It represents the case where thermal short cut and boundary resistance just cancel.

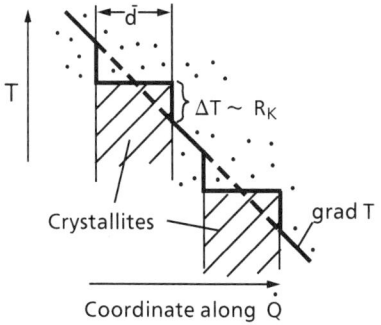

Fig. 5.5. Temperature profile in direction of the heat flux $\dot{Q}$ of an amorphous solid with crystallites embedded.

Calculation of the Interface Resistance R_K. At an interface of different materials or phases there is normally a jump in the phonon density and phonon velocity. This leads to phonon reflections and to an abrupt change in temperature ΔT which is equivalent to a thermal resistance R_K [5.11; 5.12]:

$$R_K = \left(\frac{\Delta \dot{Q}}{\Delta T}\right)^{-1} A \; ; \quad m^2 K/W \qquad (5.17)$$

where A is the interface area involved and $\Delta \dot{Q}$ is the amount of heat power flux which is reflected at each interface area. It can be calculated if the temperature dependence of $\dot{Q}$ is known. The heat or phonon energy flux $\dot{Q}$ is given by:

$$\dot{Q} = U(\omega_{max}, T, \bar{v}) \, \bar{v} \, A \, \rho \; ; \quad W \qquad (5.18)$$

where U is the internal energy which has been calculated from the density of states $D(\omega)$, the probability of states $<m>$ and the phonon energy in Section 3.2, Eq. (3.5). The procedure is exactly the same as applied for calculating the specific heat (temperature derivative of U; at low temperatures the Debye law of specific heat $c \propto T^3$ is derived from $U \propto T^4$). According to Section 3.2, the following relation holds at low temperatures

$$U = \phi(\omega_{max}, \bar{v}) \, T^4 \qquad (5.19)$$

The function ϕ contains all temperature-independent variables of U, especially the maximum frequency ω_{max} and the mean phonon velocity $\bar{v}$. Hence

$$\dot{Q} \propto T^4 \qquad (5.20)$$

which yields

$$\frac{d\dot{Q}}{dT} \approx \frac{\Delta \dot{Q}}{\Delta T} \propto T^3 \qquad (5.21)$$

Thus, looking at Eq.(5.17), the T^{-3} dependence of the interface resistance can be understood

$$R_K \propto T^{-3} \qquad (5.22)$$

This behavior of the boundary resistance is due to the large temperature dependence of U at low temperatures. At higher temperatures $U \propto T$ (Dulong - Petit range), and the interface resistance in this approximation is temperature-independent.

For an accurate calculation of the temperature jump at an interface, a transmission coefficient Γ is introduced which is different for transverse (subscript t) and longitudinal (subscript l) phonons. Evaluation of Eq.(5.21) gives in an approximation, provided that Γ is assumed to be independent of temperature.

$$R_K = \frac{15 \, \hbar^3}{\pi^2 k_B^4} \left(\frac{\Gamma_l}{v_l^2} + \frac{2\Gamma_t}{v_t^2}\right)^{-1} T^{-3} \qquad (5.23)$$

The transmission coefficients Γ mainly depend on the phonon velocities of both phases. The transmission term (in parentheses) of Eq.(5.23) is the same for the amorphous and crystalline phases. Thus, the same amount $\Delta \dot{Q}$ of phonon energy flux is reflected at the amorphous/crystalline and the crystalline/amorphous interfaces at the same temperature.

THERMAL CONDUCTIVITY

The validity of the T^{-3} dependence has been found only in special cases. It describes only the thermal resistance of the interface layer.

Boundary resistances in semicrystalline polymers are difficult to examine because the thermal conductivity of the amorphous phase plays a role. Nevertheless, the strong decrease at low temperatures of the thermal conductivity of partly crystalline polymers can be considered as confirmed. A model applicable to semicrystalline polymers with the individual thermal resistances interconnected in a network is shown in Figure 5.6. The thermal resistance is inversely proportional to the thermal conductivity. The small thermal resistance R_C of the crystallite embedded in an amorphous matrix is neglected. In this model a cubic elementary cell of size x is considered in which a quadratic crystallite of length $\bar{d}$ is embedded.

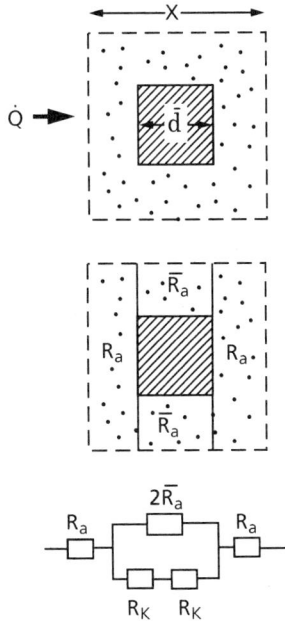

Fig. 5.6. Model for calculating the thermal conductivity of semicrystalline polymers with a parallel and serial connection of thermal resistances.

x can be calculated from the crystalline fraction f_c and from $\bar{d}$.

$$x = \bar{d} f_c^{-1/3} \qquad (5.24)$$

This gives:

$$\kappa_{semi} = \frac{1}{x}\left[2Ra + (1/\bar{R}_a + \bar{d}^2/2R_K)^{-1} \right]^{-1} \qquad (5.25)$$

with

$$2R_a = \frac{x-\bar{d}}{\kappa_a x^2} \quad \text{and} \quad \bar{R}_a = \frac{\bar{d}}{\kappa_a(x^2-\bar{d}^2)} \qquad (5.26)$$

This makes evident the influence of the crystallite size $\bar{d}$ on the thermal conductivity at low temperatures. For a given crystalline content there will be a greater surface area for smaller crystallites and hence the effect of the boundary resistance will be greater for polymers containing finer crystallites. It has been shown in several publications (e.g., Choy and Greig [5.13]) that Eq.(5.25) is consistent with the results obtained at low temperatures. However, this cannot be considered as a proof because all the transmission coefficients are not well known and thus serve solely as fitting parameters.

Information on R_K can be obtained from the crossover temperature where the thermal conductivity of the semicrystalline modification is equal to that of the amorphous phase: $\kappa_{semi} = \kappa_a$. This means that the thermal "short cut" in a crystallite is just compensated by its boundary resistance (see Fig.5.5). From Eq.(5.26) it easily can be seen that for this case

$$2R_K = \bar{d}/\kappa_a ; \quad m^2, K/W \tag{5.27}$$

This relation can be used to determine at the crossover temperature the value of R_K if the crystallite diameter $\bar{d}$ is known or vice versa. Theoretically the crossover temperature should be independent of the crystalline fraction, if it does not influence the crystallite size. As seen from Figure 5.7 for PETP, this is roughly fulfilled for a large range of crystalline fractions f_c.

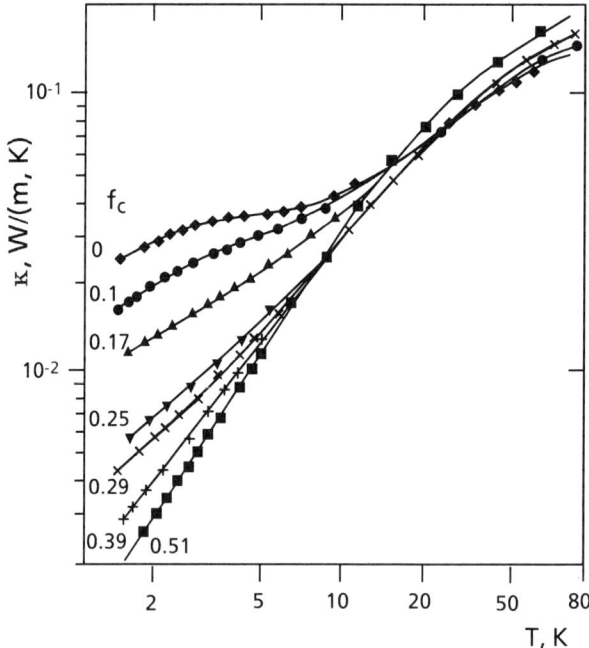

Fig. 5.7. Thermal conductivity of PETP with different crystalline fractions f_c versus temperature [5.13].

The crossover temperature, however, depends on the crystallite size of a polymer. The enormous influence of the particle size, at a constant filling, is shown in Figure 5.8. Instead of crystallites, copper particles were used which allow a defined variation of the particle diameters [5.11]. As can be seen from these curves, the crossover temperature at a constant filling varies with particle diameters. The specific surface area has been the variable in this experiment.

A similar result, but with less defined particle sizes, can be obtained from comparing Figures 5.7 and 5.9. The crossover temperature can be obtained from the intersection of thermal conductivity curves from materials with different crystalline contents f_c. In Figure 5.7, PETP has been investigated which contains relatively small crystallites. Thus, the specific surface area is relatively large. The crossover temperature lies at 15K. For comparison, HDPE is known to contain relatively large crystallites, and the crossover temperature is shifted to lower temperatures. From Figure 5.9 it can be seen that the crossover temperature is about at 2K.

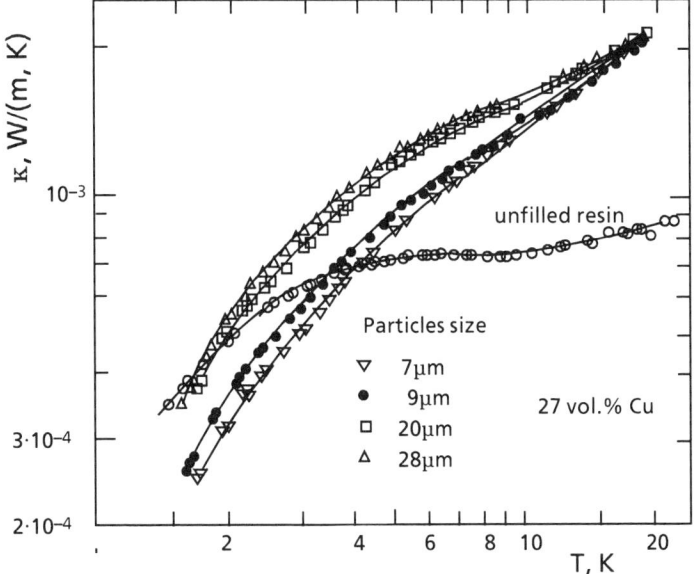

Fig. 5.8. Thermal conductivity of a pure epoxy matrix and filled with different diameter copper particles at constant filling [5.11].

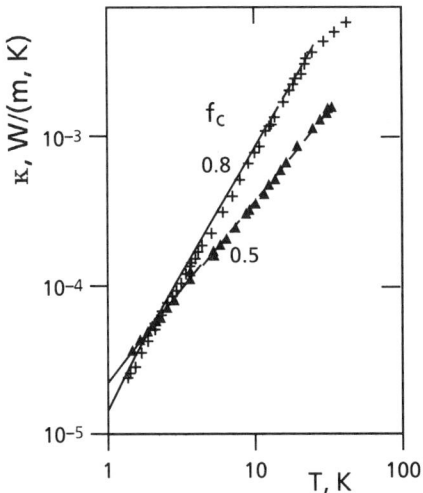

Fig. 5.9. Thermal conductivity of HDPE with two different crystalline contents f_c [5.10].

In conclusion it should be stated that the interface scattering at filler particles at low temperatures generally offers the means of generating thermal insulators with an extremely low thermal conductivity. The only decisive factor is the greatest possible interface area per filler fraction, i.e., the greatest possible specific surface. The thermal conductivity of the filler particles does not play a role. With extremely small quartz particles (0.1µm and 60vol.% filling) the thermal conductivity at 2K could be made about 50 times smaller than that of the pure polymer matrix [5.14].

Above 30K thermal conductivity is determined mainly by the conductivity $\kappa_a(T)$ of the amorphous component and the crystalline content f_c.

$$\kappa_{semi} \approx \frac{1+2f_c}{1-f_c} \kappa_a(T) \qquad (5.28)$$

5.5 RESULTS AND DISCUSSION

The thermal conductivity of several polymers is plotted in Figures 5.10a and 5.10b at different temperature ranges.

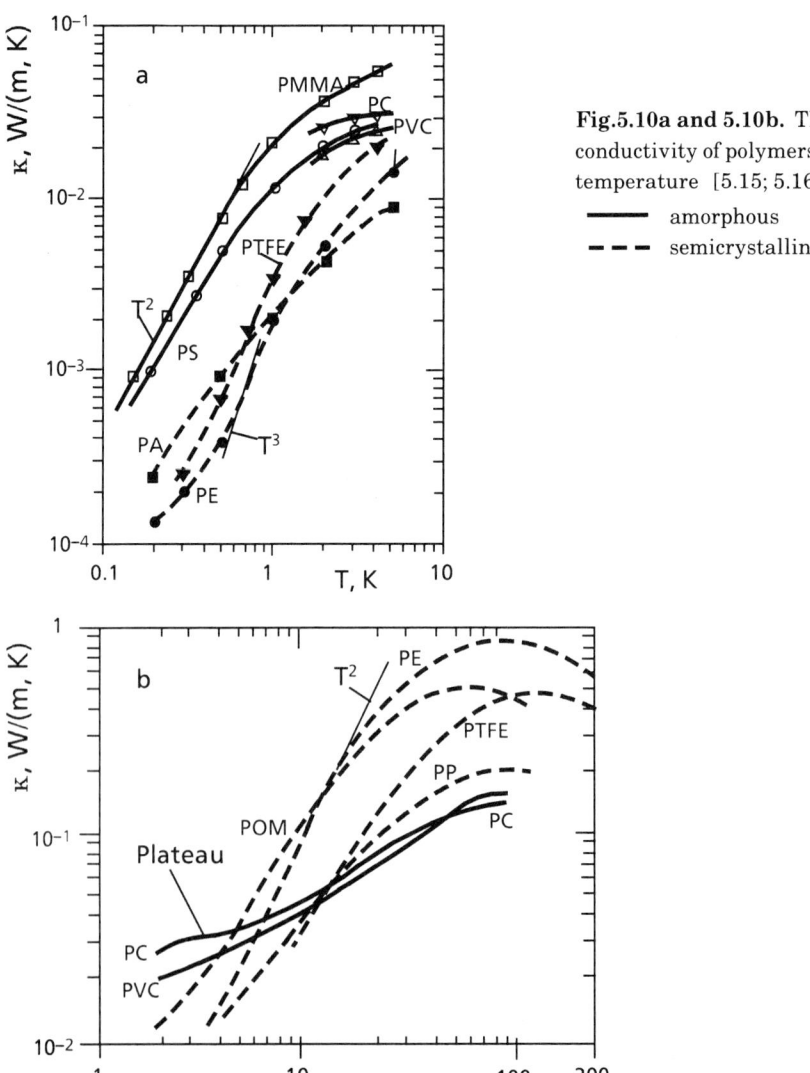

Fig.5.10a and 5.10b. Thermal conductivity of polymers versus temperature [5.15; 5.16; 5.17].

—— amorphous
- - - semicrystalline

For **amorphous polymers** below 1K a T^2 dependence can be seen which is attributed to phonon scattering at tunneling systems.

For **semicrystalline polymers** below 10K the values of κ are lower by nearly one order of magnitude than those of amorphous ones due to the boundary resistance. The temperature dependence is weaker than T^3 which would be expected from a pure interface resistance. The reason lies in the participation of the amorphous phase, which exhibits a lower temperature dependence. Roughly, a relation κ ∝ T^2 has been found for semicrystalline polymers below 10K (crystalline content about 60 vol %). Above a certain crossover temperature the high thermal conductivity of crystallites becomes dominant and semicrystalline polymers show higher values. Between 5K-15K the plateau region of amorphous polymers is indicated. It is more pronounced for epoxy resins (see Fig. 5.12). Thermal conductivity of amorphous polymers always increases with temperature. For several semicrystalline polymers it surpasses a maximum and decreases a little at increased temperatures. This can be seen from Figure 5.10b for PTFE and especially for POM.

5.6 GENERAL FEATURES OF THERMAL CONDUCTIVITY

The thermal conductivity of polymers at cyrogenic temperatures depends little on the chemical composition. For semicrystalline polymers it depends mainly on the crystalline content and on the crystallite size. For thermosets the cross-link distance is an important parameter. Thermal conductivity is determined essentially by the following structural dimensions in a polymer:

- Spacing of scattering centers (interatomic distance, chain spacing).
- Cross-link distance for thermosets.
- Crystallite size for semicrystalline polymers.

For discussion these structural lengths all are indicated by $\bar{d}$. Information on the influence of $\bar{d}$ on thermal conductivity can be obtained by comparing $\bar{d}$ with the phonon wavelengths. A useful reference length is the so-called dominant phonon wavelength λ_D[4], which is representative of phonons, carrying most of the heat at a temperature T. In analogy to optical resolution the following relations hold:

$\bar{d} \ll \lambda_D$ no resolution of a structure
$\bar{d} \approx \lambda_D$ dependence on the size $\bar{d}$
$\bar{d} \gg \lambda_D$ detail resolution within $\bar{d}$.

The first two cases are of particular importance at low temperatures. Two examples are chosen to illustrate this behavior.

---
[4] λ_D can be calculated from the Debye temperature Θ_i of a vibrational mode i, the mean spacing r_i of the oscillators and the inverse temperature.

$$\lambda_D \approx \frac{\Theta_i}{T} r_i \qquad (5.29)$$

Three representative ranges of values are given as a guideline:

1 K : $\lambda_D \approx$ 25 to 40nm
4 K : $\lambda_D \approx$ 6 to 8nm
30 K : $\lambda_D \approx$ 1.5 to 2nm

5.6.1 Cross-link Distance of Thermosets

An essential structural parameter of cross-linked polymers (thermosets) is the cross-link distance. Cross-link points act as strong phonon scattering centers, and thus are a source of thermal resistance. The smaller $\bar{d}$ is, the more cross-link points exist as strong scattering centers and the smaller the thermal conductivity becomes. The influence of the cross-link distance $\bar{d}$ depends on its relation to λ_D. The measurements suggest that there is a dependence on cross-linkage, in the low-temperature range where the dominant phonon wavelength λ_D attains the order of magnitude $\bar{d}$. The influence of the cross-link distance of epoxy resins on the thermal conductivity is shown in Figure 5.11. The cross-link distance $\bar{d}$ was varied from ≈ 1.5 to 12nm by inserting a different number of chemically identical segments between cross-link points. At 5K the dominant phonon wavelength λ_D is about 6nm, which is of the order of the range chosen for $\bar{d}$. Thus, the thermal conductivity is somewhat influenced by $\bar{d}$.

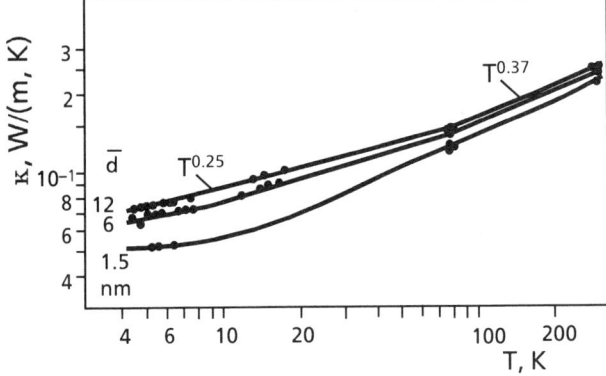

Fig. 5.11. Thermal conductivity of epoxy resins with different cross-link distances $\bar{d}$ [5.18].

At high temperatures λ_D gets small and structural scattering within the polymer chains dominates the influence of the cross-link points. As seen from Figures 5.11 and 5.12 the influence of $\bar{d}$ becomes smaller at increased temperatures.

Below 1K there is $\lambda_D \gg \bar{d}$ and cross-link points should not be resolved. This is not quite consistent with measurements shown in Figure 5.12. Rosenberg et al attributed this behavior to the fact that below 1K tunneling processes become dominant whose densities of state could depend on the cross-link distances. A different result, however, has been found by Berman [5.20]. They artificially cross-linked PS, and did not find any dependence on cross-linkage below 0.4K.

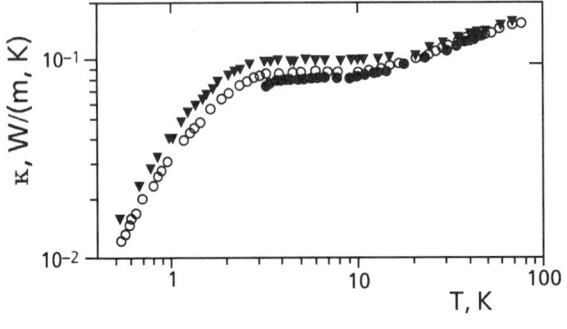

Fig. 5.12. Thermal conductivity of epoxy resins with different cross-link distances $\bar{d}$ [5.19].

5.6.2 Plateau Region of Amorphous Polymers

A peculiar feature of most amorphous polymers is the nearly constant thermal conductivity between about 5K to 15K. This can be seen from Figures 5.10b and 5.12. Another example is given in Figure 5.13 for PI (Kerimid) [5.14]. A convincing explanation is difficult. Some explanation has been given by Zaitlin and Anderson, who assume the amorphous polymers to act as low pass filters in this region (see Fig.5.2). The mean free path drops drastically with frequency and reduces the energy transfer.

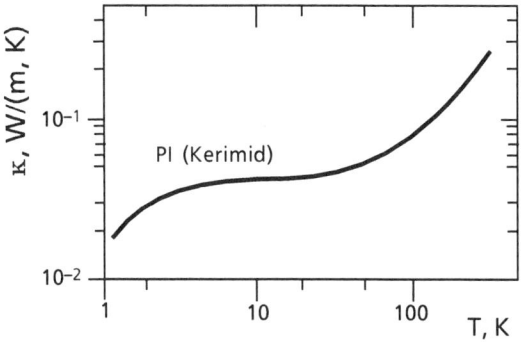

Fig. 5.13. Thermal conductivity of PI versus temperature.

5.6.3 Anisotropic Thermal Conductivity

Monocrystalline and highly oriented polymers exhibit a strong anisotropy of the thermal conductivity at high temperatures. Main heat transport occurs along the polymer chains via short wavelength phonons. Perpendicular to the chains only long wavelength phonons (wavelength larger than the chain distance) can be activated which carry less energy. Thus, anisotropic heat transport results.

At low temperatures, however, only long wavelength phonons are activated. If their dominant wavelength λ_D is larger than the interchain distance, then phonon propagation is possible in all directions, even perpendicular to the chains. Isotropic thermal conductivity results at sufficient low temperatures. This is shown in Figure 5.14 for aligned and isotropic POM.

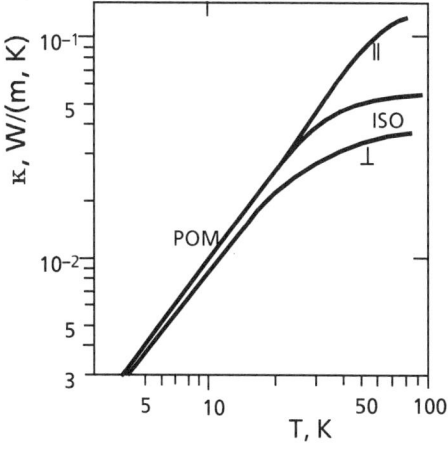

Fig. 5.14. Thermal conductivity of isotropic and aligned POM [5.16].

For an aligned polymer, $\lambda_\parallel$ gets higher and $\lambda_\perp$ smaller above $\approx 20K$. Below 20K there is practically no resolution of the anisotropic chain structure since only long wavelength phonons are excited: $\lambda_D \geq \bar{d}$ with $\bar{d}$ being the chain spacing.

5.7 THERMAL DIFFUSIVITY

The temperature equalization in a body with an inhomogeneous temperature distribution is controlled by the coefficient of thermal diffusivity a(T). It is defined by the thermal conductivity κ, the specific heat c and the density ρ.

$$a(T) = \frac{\kappa}{c\,\rho}\; ;\quad m^2/s \tag{5.30}$$

In Figure 5.15 examples are given for an amorphous and a semicrystalline polymer. At low temperatures a strong rise in the thermal diffusivity occurs owing to the strong decrease of the specific heat c. The thermal conductivity is much less dependent on temperature than the specific heat. For amorphous polymers the thermal diffusivity at 4.2K is higher by two orders of magnitude compared to RT. Semicrystalline polymers, however, show a weaker dependence on temperature. Thermal conductivity at low temperatures decreases much more because of the boundary resistance. Thus, the rise of the thermal diffusivity is smaller.

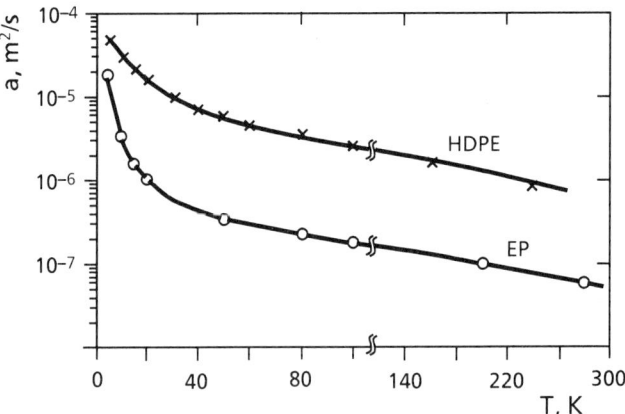

Fig. 5.15. Thermal diffusivity versus temperature

The coefficient of thermal diffusivity a(T) and the geometry of a body can be used for calculating the so-called thermal relaxation time τ_t. It is the time necessary for reducing a temperature difference ΔT exponentially to $\Delta T/e$. The thermal relaxation time τ_t generally is a function of the temperature distribution and the thermal diffusivity a(T). An example will be given for a long rod of radius r.

$$\tau_t \approx 1.1 \frac{r^2}{a}\; ;\quad s \tag{5.31}$$

For this example, a radial temperature difference ΔT within a polymeric rod of one centimeter radius, reduces to $\Delta T/e$ within a few seconds at 4.2K. At RT the value of τ_t is much larger.

5.8 SUMMARY

(1) Several phonon scattering processes make the thermal conductivity of polymers small and determine its temperature dependence. Different scattering processes work at amorphous and semicrystalline polymers.

(2) The thermal conductivity and its temperature dependence can roughly be calculated from the general relation:

$$\kappa = \frac{\rho}{3} c(T) \bar{l}(T) \bar{v}(T)$$

For amorphous polymers the mean free path $\bar{l}$ depends on both, frequency ω and temperature T. In many cases the dominant phonon approximation $\omega \propto T$ gives reasonable results. Otherwise, a more complicated relation has to be applied. For amorphous polymers only the crystalline-like portion of the specific heat can be used for calculating κ. There are localized vibrations (e.g., tunneling vibration) whose phonons increase the specific heat but do not contribute to the thermal conductivity.

(3) For semicrystalline polymers the crystallites act differently at high and low temperatures. At high temperatures crystallites increase strongly the thermal conductivity (thermal shortcut). At low temperatures the process of interface scattering at crystallites counteracts this increase. The thermal conductivity is determined by the dominance of the thermal shortcut or the interface scattering. The interface resistance is increased strongly at low temperatures: $R_K \propto T^{-3}$. Below a certain temperature depending on the crystallite size, the interface scattering reduces thermal conductivity below that of the amorphous phase. Filling of polymers with very small particles reduces thermal conductivity drastically at very low temperatures.

(4) Generally the thermal conductivity is determined by structural parameters, called $\bar{d}$ (cross-link distance, crystallite size, chain distances). The reference length is the dominant phonon wavelength λ_D.

For $\quad \bar{d} \ll \lambda_D$ no resolution.
$\bar{d} \approx \lambda_D$ strong influence of structural parameters.
$\bar{d} \gg \lambda_D$ no strong influence.

For epoxy resins the influence of the cross-link distance $\bar{d}$ on thermal conductivity is largest when $\bar{d}$ is of the order of λ_D. For aligned polymers or monocrystals the thermal anisotropy vanishes if λ_D becomes larger than the chain distance $\bar{d}$ at low temperatures: $\bar{d} < \lambda_D$. The polymer then behaves isotropically.

(5) The thermal relaxation time necessary for an equalization of an inhomogeneous temperature distribution is a function of the thermal diffusivity a. It depends on the thermal conductivity κ, the specific heat c and the density ρ. At low temperatures the value of a(T) gets very large and the thermal relaxation times become very short.

5.9 REFERENCES

5.1 Kittel, Ch.; Introduction in Solid State Physics, J. Willey & Sons, Inc. New York; (1971); p.268

5.2 Ziman, J.M.; Electrons and Phonons, Oxford, Clarendon Press; (1960); p.228.

5.3 Finlayson, D.M. and P. Mason, J. Phys. C.; (Solid State Phys.), 18 (1985); p.1791.
5.4 Phillips, W.A.; Phys. Rev. 3 (1971); p.4338.
5.5 Zaitlin, M.P. and A.C. Anderson,; Phys. Rev. B, 12 (1975); p.4475.
5.6 Farrell, D.E., J.E. de Oliveira, and H.M. Rosenberg,; Proceedings 4th. Int. Conference, Uni Stuttgart, Springer Press (1984); p.422.
5.7 Tua, P.F., S.J. Putterman, and R. Orbach,; Phys. Lett. 98 A (1983); p.357.
5.8 Reese, W. and J.E.Tucker,; J. Chem. Phys. 43 (1965); p.105.
5.9 Engeln, J. and M. Meissner; in "Nonmetallic Materials and Composites at Low Temperatures"; Vol. 2; Plenum Press; (1980); p. 14.
5.10 Kolough, R.J. and R.G. Brown,: J. Appl. Phys. 39 (1968); p.3999.
5.11 Schmidt, C.; Cryogenics, Jan. 1975; p. 17.
5.12 Andersen, A.C. and R.B. Rauch; J. Appl. Phys.; Vol. 11 (1970); p. 3648 and R.E.Peterson, and A.C. Anderson,; J. Low Temp. Phys. II (1973); p.639.
5.13 Choy, C.L. and D. Greig,; J. Phys. C., Solid State Phys. 8 (1975); p.3121.
5.14 Claudet, G., F. Disdier, and M. Locatelli; in "Nonmetallic Materials and Composites at Low Temperatures", Vol. 2, p. 131, Plenum Press (1979).
5.15 Greig,D. and M.Sahota; in "Nonmetallic Materials and Composites at Low Temperatures" Vol. 3,p.9. Eds.: Hartwig, G., Evans, D.;Plenum Press; New York (1986)
5.16 Choy, C.L. and D. Greig, ; J. Phys. C., Solid State Phys. 10 (1977); p.169.
5.17 Reese, W.; J. Appl. Phys. 37 (1966); p.864.
5.18 Hartwig, G.; Progr. Colloid + Polymer Sci. 64 (1978); p.56
5.19 Kelham, S. and H.M. Rosenberg,; J. Phys. C., Solid State Phys. 14 (1981); p.1737.
5.20 Berman, B.L., R.P. Madding, and J.R. Dillinger,; Phys. Rev. Lett., 30 A (1969); p.315

MOLECULAR PLACE CHANGES AND MECHANICAL DAMPING SPECTRA

Contents

6.	Molecular place changes and mechanical damping spectra	119
6.1	General survey	119
6.2	Nomenclature of glass transitions	120
6.3	Single place changes and potential barriers	121
6.4	Results and discussion	124
6.5	Theory (barrier jumping, tunneling processes)	127
6.6	Asymptotic behavior at low temperatures	136
6.7	Summary	137
6.8	References	138
Appendix		
6 A	Data	

6 MOLECULAR PLACE CHANGES AND DAMPING SPECTRA

6.1 GENERAL SURVEY

The loosely packed structure of amorphous polymers makes it possible that, for small segments or side groups, two (or more) neighboring potential minima exist which are separated by a small potential barrier $\Delta\phi$. Place changes between two potential minima take time, the so-called relaxation time τ. It is a function of the potential distribution and the temperature. By external mechanical or dielectric loading the populations of the double well minima are disturbed (see Fig.6.10). The equilibrium is restored by place changes. As the relaxation time of place changes is finite, there is a delay between loading and deformation of a material, which causes dissipation of the loading energy. For cyclic loading a loss-angle δ occurs between stress and strain, which is a function of the temperature and of the load frequency. It determines the damping behavior.

Damping spectra can be used as indicators of place changes. The damping behavior of amorphous polymers is plotted schematically versus the temperature in Figure 6.1. The property shown is the mechanical loss-factor $\tan\delta$, which can be derived from the logarithmic decrement of damped oscillations, from sound absorption and from hysteresis measurements (see Section 6.4).

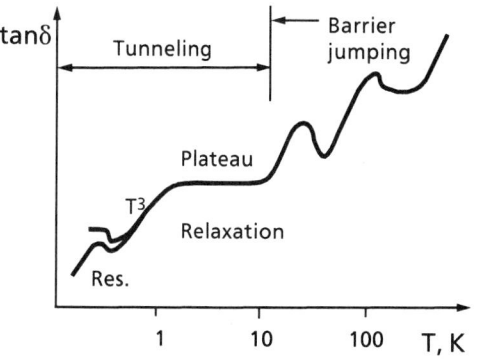

Fig. 6.1. Loss-factor $\tan\delta$ of amorphous polymers plotted schematically versus temperature at a constant frequency.

Jumping over Potential Barriers. Above 20K damping spectra show maxima which are attributed to singular barrier jumping processes. The corresponding relaxation time can be calculated with the Arrhenius equation which is based on Boltzmann statistics. In the temperature range of 30K-300K place changes by singular mobilities of certain chain segments or side groups prevail. Singular barrier jumping is correlated to a damping maximum, which characterizes a low-temperature glass transition. The position of a damping peak depends on the temperature and the measuring frequency. In many cases it is possible to determine the potential barrier height $\Delta\phi$ and the relaxation time experimentally from the temperature-frequency position of the damping maxima. Sections 6.4 and 6.5 are devoted to potential barriers and the correlated damping behavior. It should be mentioned that, like other properties, the damping behavior might be changed by the water content. Especially damping peaks in the vicinity of 170K (at 10Hz) caused by main chain mobilities might be different for dried and nondried polymers (e.g., PEI).

Tunneling Processes (dominant below 15K). Tunneling processes show a different damping behavior. They are of quantum-mechanical nature, and describe place changes via tunneling through potential barriers. Those processes exhibit a much smaller relaxation time than jumping processes at very low temperatures and a given potential barrier. Therefore, tunneling processes dominate at very low temperatures. The relaxation time is given by the quantum-mechanical lifetime of an excitation. It depends on temperature and several parameters which characterize the potential barrier. Two types of processes are known: resonant and relaxation tunneling. They are dominant in different frequency- or temperature ranges. A plateau region below 15K and a steep decay at very low temperatures are characteristic of tunneling damping processes. This will be explained in Section 6.5.2. At very low temperatures and high frequencies resonant tunneling can be excited.

Place changes are not only combined with dissipation processes, they also influence several properties. (The influence of place changes on properties has already been outlined in Fig. 1.8). A stepwise change of moduli and permittivity occurs at the onset of singular barrier jumping processes. Tunneling processes effect only a smooth property change.

6.2 NOMENCLATURE OF GLASS TRANSITIONS

The onset of place changes at a relatively high transition rate can be defined as a glass transition. It is accompanied by a damping maximum. Sometimes it is difficult to find appropriate criteria for defining primary and secondary glass transitions. This is especially true for semicrystalline polymers. Customarily, the damping maxima are labeled by Greek letters α, β, γ with decreasing temperatures. The main glass transition is denoted by α. For many semicrystalline polymers the α transition is related to the crystalline phase; the first transition of the amorphous phase in many cases is denoted by β or γ.[1] It signifies the onset of segmental chain mobilities in the amorphous domains.

Throughout this text book a different nomenclature is used, which is related to the type of molecular motions and which conveys more information about glass transitions:

---

[1] It should be mentioned that for many semicrystalline polymers the secondary glass transition of the amorphous phase is connected with a much larger modulus step than that at the main glass transitions.

Primary glass transition: main glass transition; flexibility of the whole chain;

Secondary glass transition: onset of segmental mobilities in the molecular chains, e.g., crankshaft motions;

Tertiary glass transition: onset of side group mobilities.

In most cases this notation is sufficient. For polymers with several secondary or tertiary transitions, the temperature-frequency position of the damping maxima may be added.

6.3 SINGLE PLACE CHANGES AND POTENTIAL BARRIERS

Below RT it is almost sufficient to concentrate on secondary or tertiary glass transitions. Single (or local) place changes means the mobility of a certain molecular unit. Since the related potential distributions are not exactly the same within a polymer, the damping spectra are slightly smeared out. At low temperatures, only low energy barriers allow molecular groups to change places. Rotations around a C-C or C-O binding axis are the most important low-temperature mobilities. They are *a priori* free and without barriers. Barriers, however, arise from interactions with neighboring chains (interchain interaction, steric hindrance) and from interactions of side substituents of a chain backbone (intrachain interaction). One well-known example is the intrachain rotational potential of a CH_3 end group whose principal origin can be seen in Figure 6.2.

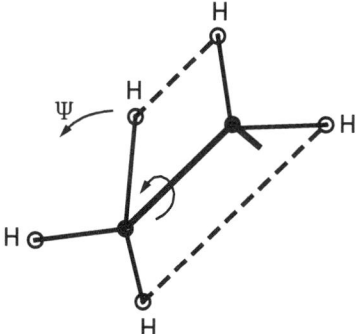

Fig. 6.2. Intrachain interaction of a CH_3 and a CH_2 monomer.

The potential barriers arise from secondary binding forces of the H atoms of the neighboring monomers (dashed lines). Rotation of these monomers in this example causes symmetric barriers and potential minima, as shown in Figure 6.3. For most cases the distribution is less regular.

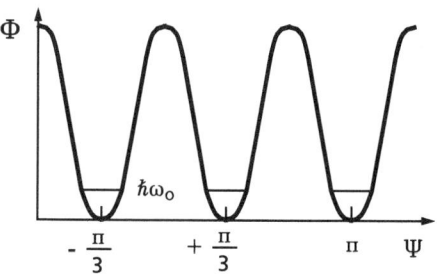

Fig. 6.3. Rotational potential of a CH_3 end group versus the azimuthal angle ψ.

6.3.1 Mobilities within Chains

Crankskaft Motions. Within a long zigzag chain a single monomer cannot perform rotations. But, a collective motion of several monomers, like a crankshaft, might be possible. With respect to the valenceangle, at least 4 or 5 monomers are necessary to allow a rotation around a common axis (see Fig. 6.4).

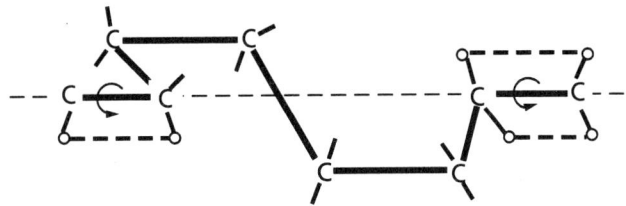

Fig. 6.4. Crankshaft motion in a homogeneous chain.

Crankshaft motions are controlled by medium-high potential barriers, which are caused mainly by intrachain interactions of side atoms (dashed lines in Fig. 6.4) and by interchain interactions. Schatzki by potential theory calculated the potential barrier of crankshaft motions with 5 monomers in PE [6.1]. The calculated value of 54kJ/mole is consistent with experimental results (50-60kJ/mole). Those potential barriers control low-frequency relaxations in the medium temperature range from 150K to 250K.

In Section 6.4 crankshaft motions will be studied on a series of polymers with different dipole moments and sizes of the side atoms of the chain backbone. Inter- and intrachain interactions of the crankshaft segments increase by increasing sizes and dipole moments of side atoms, and higher potential barriers result.

Influence of Oxygen Bridges. The Van der Waals forces between side atoms of sucessive monomers can be reduced by spacers, e.g., by oxygen bridges in the backbone of molecules. As an example PE and POM are compared.

$$\text{PE: } -CH_2-CH_2- \quad \text{POM: } -O-CH_2-O-CH_2-$$

The forces between H atoms of POM are smaller than of PE, and therefore a smaller potential barrier is expected for crankshaft motions of POM. The contrary has been found from experiments. At a constant frequency $f \approx 6Hz$, the temperature positions of the damping peaks are for PE: 158K and for POM: 213K (see Figs. 6.9 and 4.15). The reason is that oxygen atoms, which possess a large dipole moment, are moved with the crankshaft segment, thus increasing the interaction forces with neighboring chains.

Rotation of Aromatic Rings. Another possible collective motion of chain segments is known for aromatic rings. An axis of rotation is possible at certain ring configurations. For PEEK molecules, for example, there are aromatic rings, which are "pivoted" in oxygen bridges at both ends (see Appendix 1A).

In this case, the oxygen bridges are "fixed" spacers which reduce only the interchain forces between the relevant side atoms (dashed lines). The potential barrier should be small, but only a shoulder occurs for PEEK at very low temperatures, namely at about $\approx 35K$. There is an overlap from other glass transitions.

6.3.2 Mobilities of Side Groups

Wagging of Phenyl Rings in PS. Wagging, in this case, means small rotations of the phenyl ring with the chain segment as the axis (see Fig. 6.5a). When two rings of neighboring chains are wagging in close vicinity, their Van der Waals interaction (dashed lines) builds up a small potential barrier, which is superimposed on the strong rotational potential. Wagging is then correlated to a jumping process. The potential distribution has been calculated by Wada [6.2]. His results are plotted in Figure 6.5b. A mean potential barrier of 10kJ/mole has been calculated which is consistent with experimental results. A damping maximum exists at 40K and 5Hz. Wada considered in addition rotations of ring planes. The barrier of this motion is smaller. The mechanical damping spectra, however, did not show an activation of this type of motion. (If the rings are made asymmetric, e.g., by a substitution with chloride, a dielectric activation of ring plane rotations is observed [6.2]).

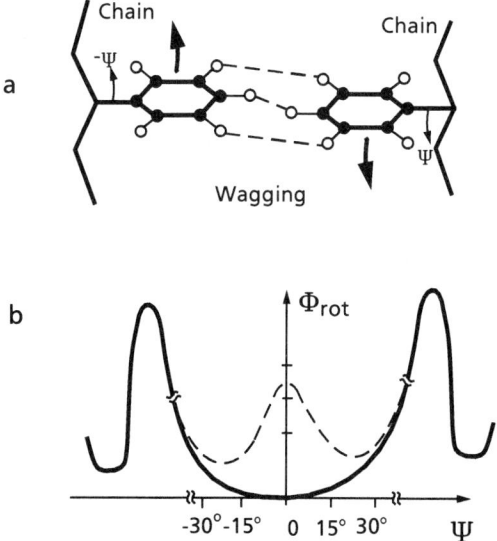

Fig. 6.5. a) Wagging of two phenyl rings. b) Potential distribution of wagging phenyl rings.

Rotations in Side Groups of Polyacrylates. Polyacrylates with different lengths of side groups exhibit damping peaks, which are due to different segmental mobilities. They are sketched in Figure 6.6. The results are plotted in Fig. 6.7.

6.4 RESULTS AND DISCUSSION

The concept of molecular mobilities in amorphous polymers is subject to many uncertainties. A quantitative treatment is difficult. Before entering into theoretical considerations, it will be demonstrated in this section that the experimental results are qualitatively consistent with the conceptions of single molecular mobilities. The results concentrate on damping maxima in secondary and tertiary glass transitions in the low-frequency region (5-15Hz). The low-frequency measurements were performed with a torsion pendulum. Other methods at various frequency ranges are compiled in the foot note [2].

Investigations of homologous series of polymers are made which show special trends or tendencies of molecular mobilities. The determination of potential barriers from damping spectra will be treated in Section 6.5.

6.4.1 Polyacrylates with Different Lengths of Side Groups

A homologous series of polyacrylates with different lengths of their side groups is shown in Figure 6.6. PMMA has the smallest side group. A methyl group is connected with the carbonyl group. For PEMA it is assumed that the ethylene group has a rotational mobility around the oxygen bond with a very low potential barrier. A barrier height of 9kJ/mole has been calculated by Wada which is consistent with experimental results [6.3; 6.4]. The assumed axes of rotations are denoted by arrows. The damping spectra of these polyacrylates are shown in Figure 6.7.

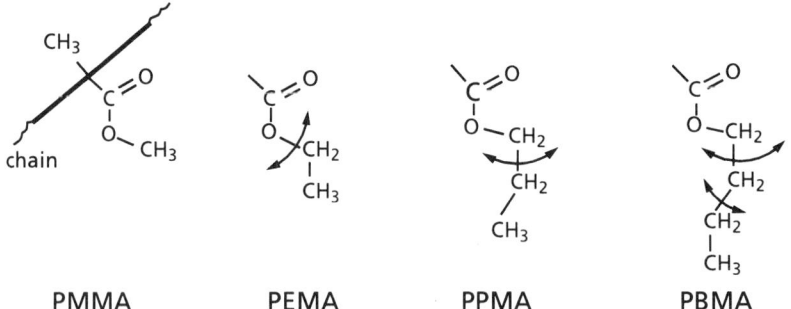

Fig. 6.6. Polyacrylates with different side groups.

---

[2] Depending on the frequency range, different measuring methods are applied to measure the loss-angle δ:
- below 10^{-1}Hz: creep measurements;
- from 0.1Hz to 50Hz: decay of amplitudes of free oscillations, e.g., in a torsion pendulum. The decisive parameter is the logarithmic decrement $\Lambda = \ln(A_1/A_2)$ of sucessive amplitudes A_2 and A_1. It holds: $\tan\delta = \Lambda\pi$
- below 10Hz: hysteresis effect of stress-strain curves;
- kHz range: forced oscillations measured by the vibrating reed method;
- MHz range: hysteresis effect from forced oscillations in ultrasonic measurements and ultrasonic absorption measurements. The sound energy absorption coefficient a_s is correlated to the loss-factor by: $\tan\delta = a_s \lambda \pi$, where λ is the wavelength of the sound.

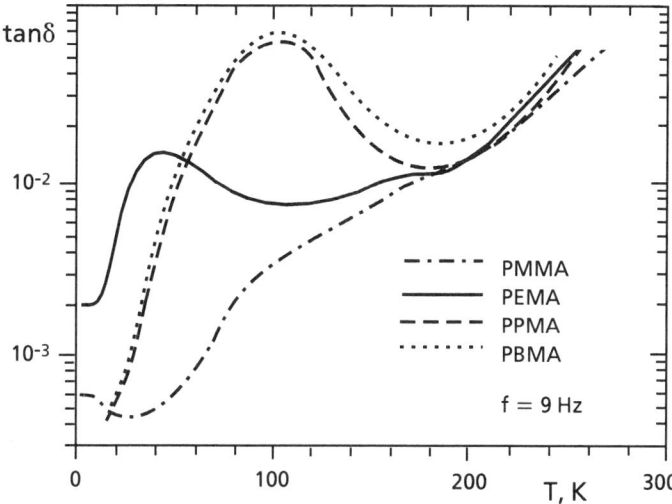

Fig. 6.7. Damping spectra of polyacrylates in the low-frequency region versus temperature [6.4].

PMMA has the smallest side group, and a damping peak would be expected at a relatively low temperature. Instead, a damping minimum has been observed around 30K. There is a small increase in tanδ with decreasing temperatures; below 10K it opens into a plateau region, which arises from tunneling processes (see Fig. 6.16). Maybe, a peak is buried in the plateau region. An explanation of this strange behavior will be given in Section 6.5.2. PEMA has a side group, which is longer by a CH_2 monomer. A damping peak occurs at 40K. PPMA has an even longer side group. The CH_2-CH_3 group is again mobile. It is, however, connected to a CH_2 monomer and thus makes more intrachain interactions than in PEMA where it is bonded to an oxygen bridge. The potential barrier is higher and a damping maximum occurs at 100K. The same situation is true for PBMA and all other polyacrylates, namely the rotation of the same CH_2-CH_3 group around a different number of CH_2 monomers. As seen in Figure 6.7, a damping peak occurs at nearly the same temperature (100K) for those polyacrylates.

6.4.2 Dependence of Crankshaft Motions on Dipole Moment and Size of Side Atoms

If the dipole moment and the size of side atoms of a molecular backbone and their dipole moments are increased, the steric hindrance and the intrachain forces and, thus, the potential barriers are increased. This is demonstrated on a series of homologous polymers with side atoms of increased sizes and dipole moments [6.5].

Polymers with H, F and CL Side Atoms. The following series is considered:

$$\begin{bmatrix} H & H \\ -C-C- \\ H & H \end{bmatrix}_n \quad \begin{bmatrix} F & F \\ -C-C- \\ F & F \end{bmatrix}_n \quad \begin{bmatrix} H & H \\ -C-C- \\ Cl & H \end{bmatrix}_n \quad \begin{bmatrix} H & H & H & H \\ -C-C-C-C- \\ Cl & Cl & H & Cl \end{bmatrix}_n$$

HDPE 　　　　 PTFE 　　　　　　 PVC 　　　　　　　　 PVCC

The size of side atoms of a backbone can be characterized by the bond distance plus the atomic radius (convergence radius):

C - H : 0.14nm
C - F : 0.21nm
C - Cl : 0.28nm

H side atoms are small and rather nonpolar, whereas chloride side atoms are large and polar. The damping peak occurs at the lowest temperature for HDPE and at the highest one for post-chlorided PVC (PVCC). For fluorine side atoms the barriers and the damping peaks lie in between. This indicates that potential barriers are increased by increased sizes and dipole moments of side atoms. The related potential barriers $\Delta\phi$ are added in Figure 6.8.

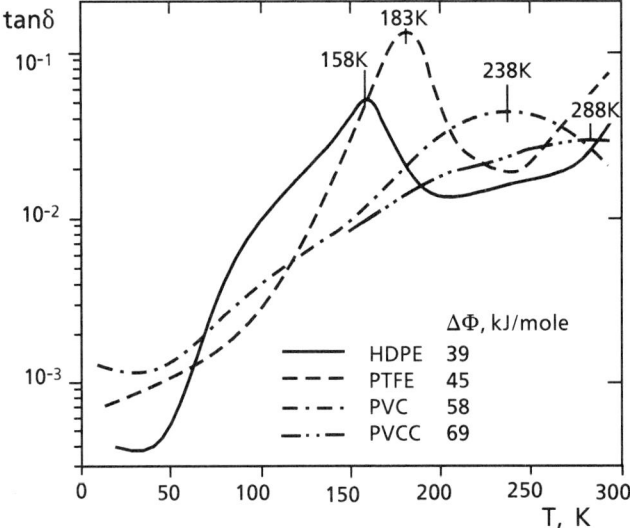

Fig. 6.8. Damping spectra of polymers having different dipole moments and sizes of side atoms. (frequency: 5 - 8Hz). The potential barriers $\Delta\phi$, calculated from Eq.(6.5), are included [6.5].

Polymers with an Increased Concentration of F Atoms, Starting from HDPE. The following series is considered:

$$\begin{bmatrix} H & H \\ -C-C- \\ H & H \end{bmatrix}_n \quad \begin{bmatrix} F & F \\ -C-C-C-C- \\ F & F & H & H \end{bmatrix}_n \quad \begin{bmatrix} F & H \\ -C-C- \\ F & H \end{bmatrix}_n \quad \begin{bmatrix} F & F \\ -C-C- \\ F & F \end{bmatrix}_n$$

HDPE   ETFE   PVDF   PTFE

The nonpolar H atoms are substituted by F atoms, which are larger and have a higher dipole moment. As shown in Figure 6.9, at a nearly constant experimental frequency, the damping maxima are shifted to higher temperatures when the concentration of F atoms is increased. However, a discrepancy is observed for PVDF, which has the same fluorine content as ETFE, but a peak occuring at an unexpectedly high temperature.

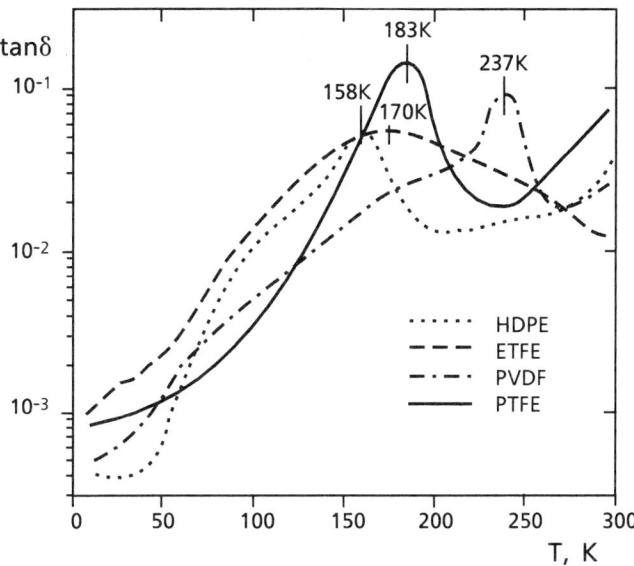

Fig. 6.9. Damping spectra of polymers with an increasing concentration of F atoms (5-8Hz).

The results discussed in this section roughly support the model of local place changes of single molecular units. The next question relates to the determination of the barrier height $\Delta\phi$. It will be shown in Section 6.5.1 that this is possible if the frequency-temperature position of a damping maximum is known.

6.5 THEORY

The equilibrium between two possible conformations of a molecular group can be disturbed by an external load. The equilibrium is restored by jumping over the potential barrier or by tunneling. Both processes are of a statistical nature; the rearrangement takes a characteristic time, the relaxation time τ. The amount of damping depends on the relation of the loading time Δt and the relaxation time τ.

For $\Delta t \ll \tau$, there is no time available for the statistical process of place changes. For $\Delta t \gg \tau$ the rearrangement directly follows the load rate. A maximum of dissipated power exists if τ is on the order of Δt. In this case, the jumping rate is equal to the loading rate. The exact correlation between Δt and τ depends on the time profile of loading. A very important type of loading is the sinusoidal loading, whose time dependence is characterized by the angular frequency ω. The periodic changes of place can then be considered as a damped oscillator with a relaxation time τ. If under sinusoidal load at an experimental angular frequency ω, the loss-factor is proportional to the following formula (see Section 7.6.2)

$$\tan\delta \propto \frac{\omega\tau}{1+(\omega\tau)^2} \tag{6.1}$$

A maximum of $\tan\delta$ exists at

$$\omega\tau = 1 \tag{6.2}$$

This general relation holds for jumping and tunneling processes. Singular maxima of tanδ, however, have been observed for jumping processes only. For tunneling processes many maxima of tanδ are overlapped and a plateau region results. The relaxation time τ is a function of the temperature T and of the potential distribution. These correlations will be considered for jumping and tunneling processes.

6.5.1 Jumping over Potential Barriers

The analytical treatment starts with a double potential with a potential barrier Δφ as shown in Figure 6.10. The abscissa can be the distance of interacting units or the azimuthal angle ψ of rotations (see Fig. 6.2).

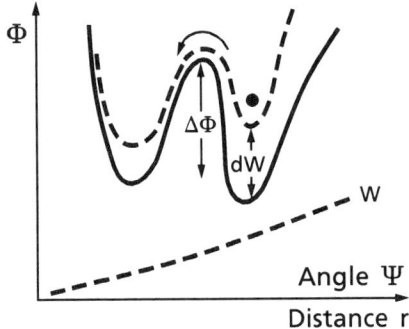

Fig. 6.10. Double well potential versus a length coordinate r or the rotational angle ψ.

——— without load energy
– – – with a disturbance dW by an external load energy W.

Without external load, an equilibrium state exists between the population of segments or side groups in both potential minima. If an external load W is applied, the distribution is disturbed and jumping processes occur when the thermal energy of candidates for jumping is equal to or higher than Δφ. The rate is proportional to the Boltzmann factor $\exp(-\Delta\phi/RT)$. The jump rate depends on this factor and on a "trying" frequency v_0, which is of the order of thermal frequencies. The relaxation time τ is the inverse jump rate and is given by the Arrhenius equation [3] [6.6]

$$\tau(T, \Delta\phi) = v_0^{-1} \exp(\Delta\phi/RT) \tag{6.3}$$

R = 8.3 J/(mole·K) is the gas constant; $v_0 \approx 10^{12}$ to 10^{13} Hz (it depends to some extent on the type of molecular motion).

In most cases one measures damping spectra as a function of the temperature at a constant frequency or vice versa. Combining of Eqs.(6.2) and (6.3) yields the following relation, which is valid at a damping maximum

$$\omega_m v_0^{-1} \exp(\Delta\phi/RT_m) = 1 \tag{6.4}$$

---

[3] It is assumed that the distortions are small compared to Δφ and that they are linearly distributed along the potential. With these approximations made, τ is independent of the load amplitude.

ω_m and T_m are the values at the damping maximum. They can be determined experimentally and used for calculating $\Delta\phi$ by Eq.(6.4). A plot of $\ln \omega_m$ versus $1/T_m$ yields a straight line whose slope is equal to $\Delta\phi$. This is shown in Figure 6.11 for different values of $\Delta\phi$. In this figure a constant value of v_0 has been assumed. Consequently, all lines meet at one point for $1/T_m \to 0$. If v_0 is not exactly known, one can determine two or more frequency-temperature pairs ω_{mi}; T_{mi}. Again, a straight line is obtained whose slope determines $\Delta\phi$.

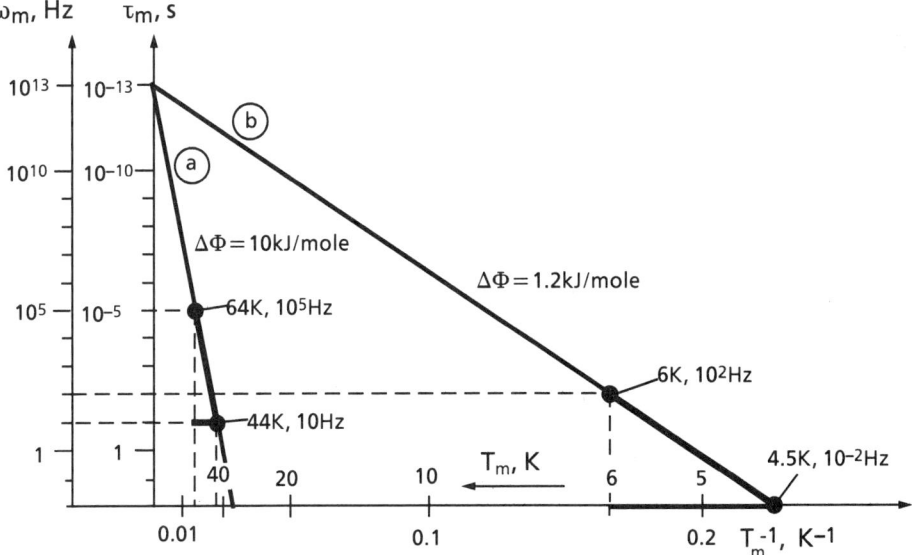

Fig. 6.11. Arrhenius plot: $\ln \omega_m$ versus $1/T_m$ for different $\Delta\phi$. τ_m is the relaxation time.

The Arrhenius equation is also valid for describing the temperature - frequency shift of secondary and tertiary glass transitions.[4] The following modification of Eq.(6.4) is easier to handle

$$T_m = (0.25 - 0.019 \log f_m)^{-1} \Delta\phi \quad (6.5)$$

The angular frequency has been substituted by the ordinary frequency f, which can be directly used, e.g., in pendulum measurements. The values in Eq.(6.5) are calculated by putting: $v_0 = 10^{13}$ Hz.

One special case, considered by Heijbour, is given for measurements made at $f_m = 1$ Hz [6.4]. From Eq.(6.5) it follows that

$$\Delta\phi = 0.25 \, T_m \quad ; \quad kJ/mole \quad (6.6)$$

This means that at one Hertz the potential barrier is directly proportional to the temperature T_m at a damping peak.

---

[4] The temperature-frequency shift of primary glass transitions is given by the WLF equation which is not a topic of this book.

Some remarks on the low-temperature relaxation time should be added for demonstrating the strong influence of $\Delta\phi$ and T on τ. At low temperatures a finite τ exists only for small $\Delta\phi$. This is demonstrated by two numerical examples, which are calculated from Eq.(6.3):

1) At 4K: $\tau = 1$ for $\Delta\phi = 1$kJ/mole
$\tau = 10^{13}$ for $\Delta\phi = 2$kJ/mole

2) The tertiary glass transitions of PS and PEMA are both characterized by a barrier of $\Delta\phi \approx 10$kJ/mole. The relaxation time is $\tau = 0.2$s at 40K and would become $\tau = 10^{110}$s at 4K.

It is characteristic of the Arrhenius relation because of the exponential that a drastic change of τ is caused by a small temperature variation. From Eq.(6.2) the similar statement can be derived that a large change of frequency ω causes only a small temperature shift of a peak. This is demonstrated by the two Arrhenius lines plotted in Figure 6.11. A four-decade frequency variation causes a temperature shift from 44K to 64K (line (a)) and from 4.5K to 6K (hypothetical line (b).) Another example is given for PS. Its low-temperature damping peak occurring ,for example, at 40K for 10Hz is shifted by only 1K at double the frequency. This feature is helpful, e.g., for pendulum measurements in which it is difficult to keep the frequency constant at different temperatures.

6.5.2 Tunneling Processes

At very low temperatures the propability of jumping processes becomes very small even for low potential barriers. Tunneling processes then become dominant. Their damping spectra are smeared out and a plateau region exists. The reason is that several parameters with a broad distribution each determine the tunneling behavior. At the low-temperature end of the plateau a strong decrease of the damping spectrum (loss-factor) occurs. There are two types of processes:

- **Resonant tunneling.**
- **Relaxation tunneling.**

The quantum-mechanical treatment of tunneling processes and their damping or relaxation behavior will be briefly outlined below. It is well known that a quantum-mechanical wave function of a particle penetrates a potential barrier [6.7]. If the barrier is sufficiently thin, then transmission of particles is possible, even if the energy conservation law is violated within a short distance. This follows from the quantum-mechanical uncertainty principle.

A double potential with a barrier of thickness d and height $\Delta\phi$ is assumed. Both potential minima differ by Δ (see Fig. 6.12). The wave functions ψ_l and ψ_r of particles in the left and right potential minimum overlap if d is small enough. This leads to mutual influencing by both neighboring potentials, which are, for example, possible positions of a molecular group. This interaction can be described by a splitting $\pm E/2$ of the energy levels of both potentials. Since higher excitations are unlikely at very low temperatures, only the ground levels $\hbar\omega_0/2$ are involved and split.

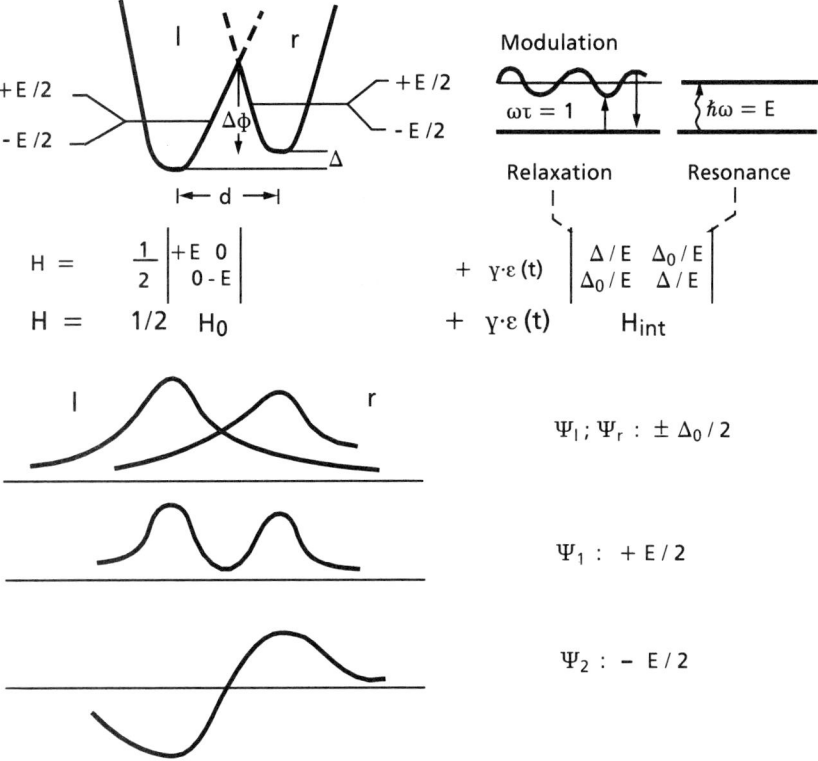

Fig. 6.12. Double well potential; the correlated wave functions, the matrix elements and the resonant and relaxation processes occurring under external loading.

The energy E can be calculated from the orthogonal wave functions ψ_1 and ψ_2. The correlated matrix $|H_0|$ has two eigenvalues $\pm E/2$, which are the energy levels. In equilibrium, a small molecular group has a certain probability of staying at one of these energy levels. A tunneling system is therefore called a "two-level system." [6.8; 6.9]. The value of E is given by

$$E = \pm \frac{1}{2}(\Delta^2 + \Delta_0^2)^{1/2} \qquad (6.7\,a)$$

with the splitting parameter $\Delta_0 = \hbar\,\omega_0 e^{-\lambda}$, where ω_0 is the frequency of the zero-point energy $\hbar\omega_0/2$. The tunneling parameter λ is determined mainly from the barrier height $\Delta\phi$ and the width d

$$\lambda = (2m\,(\Delta\phi - E))^{1/2}\,d/\hbar \approx (2m\,\Delta\phi)^{1/2}\,d/\hbar \qquad \text{for } \Delta\phi \gg E \qquad (6.7\,b)$$

where m is the mass of the tunneling particle.

Tunneling between two potential minima of a particle is described by a transition between the two energy levels $\pm E/2$. A transition can be excited by an external load, e.g., by an external strain $\varepsilon(t)$. The interaction of the external excitation with the two-level tunneling system is given by the matrix $|H|_{int}$ and the coupling parameter γ (see Fig. 6.12). When ultrasonic waves are used to excite a transition, they can be absorbed without disturbing the energy levels or they can modulate them. The first type of process is called resonant tunneling, the second relaxation tunneling. Both processes can be described by the non - diagonal and diagonal matrix elements, respectively, of $|H|_{int}$.

Resonant Tunneling Processes. These processes are characterized by the fact that the energy levels are not modulated by external loading. Only phonons are absorbed (or reemitted) which have an energy $\hbar\omega$ equal to the energy level separation E.[5] By those absorption processes the population of the ground level is shifted to the upper level. After a certain lifetime in the excited level, phonons are emitted in a random direction. Thus, an incoming phonon beam is absorbed by delayed scattering at tunneling systems. The absorption of phonons may lead to saturation effects if the intensity of the incoming phonon exceeds a critical value I_c. There is less absorption if one of the two energy levels is filled up at a faster rate than the retransition rate [6.10]. The loss-factor is thus a function of the intensity I

$$\tan\delta_{res} \propto \frac{1}{(1-I/I_c)^{-1/2}} \tanh(\hbar\omega/2k_B T) \qquad (6.8)$$

where ω is the frequency of incoming phonons (or photons upon dielectric activation).

At low frequencies resonant absorption occurs at tunneling systems with a small splitting energy E. Those processes, however, can be excited at extremely low temperatures only. At very high frequencies (10GHz), however, resonant tunneling is shifted to reasonable temperatures. Thus, very high frequencies or extremely low temperatures are the domains of resonant tunneling. (High-frequency activation is possible by electromagnetic waves and will be treated in Section 8.5.2).

Relaxation Tunneling Processes. These processes are characterized by a modulation of the energy levels through external loading. This is produced by low energy phonons; e.g., by sound waves of low frequencies. At a given temperature relaxation tunneling processes are excited by phonons of much lower frequency than required for resonant tunneling absorption. In the example given in Figure 6.12 the upper energy level is modulated by a sinewave which disturbs the equilibrium. During the period of an increased energy level transitions from the upper to the lower energy level are induced, and vice versa for the period of a decreased energy level. These transitions take some time and are therefore dissipative. The tunneling relaxation time τ_{rel} can be calculated from the diagonal matrix elements of $|H|_{ext}$ for a modulating phonon [6.11]

$$\tau_{rel} = \text{const } E^{-3}(E/\Delta_0)^2 \tanh(E/2k_B T) \qquad (6.9)$$

---

[5] This process also includes intermediate excitations to higher levels and reemission to the upper level of the tunneling systems.

The constant is mainly a temperature-independent function of the sound velocity and coupling parameters. The function $\tanh(E/2k_BT)$ is rather temperature independent in the range $E \geq 2k_BT$ considered at low temperatures. Thus, the tunneling relaxation time τ should be rather independent of the temperature if E and Δ_0 are fixed values. This behavior is very different from that of jumping processes. For single jumping processes according to the Arrhenius relation, the relaxation time at a temperature T is given by the potential barrier $\Delta\phi$ only.

Arrhenius relation: $\tau = \tau(T, \Delta\phi)$.

For tunneling processes, the barrier width d and the difference Δ of the potential minima are additional decisive parameters. (According to Eq.(6.7) they are incorporated in the parameters Δ_0 and E).

Tunneling processes: $\tau = \tau(T, \Delta_0, E)$.

The same relaxation time τ can be obtained from different tunneling systems with different combinations of the parameters $\Delta\phi$, Δ, d. The temperature dependence of the relaxation time is plotted in Figure 6.13 for jumping and tunneling processes.

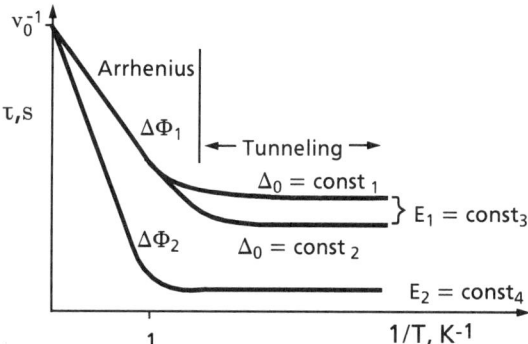

Fig. 6.13. Relaxation time τ versus 1/T according to the Arrhenius relation and due to tunneling processes with fixed parameters E and Δ_0.

The next step will be to explain the mechanical, low-temperature damping behavior arising from tunneling processes. As already mentioned, a striking feature of damping below 15K is the temperature independent plateau region at a given frequency. At first glance, its explanation seems to be easy when considering that the relaxation time τ_{rel} is nearly temperature independent at low temperatures. The situation, however, is more complicated. Not only the relaxation time of tunneling systems, but also their relative contributions are decisive. There is a large variety of different tunneling systems with the same τ, but they significantly contribute at different temperatures. This is schematically shown in Figure 6.14. Many damping peaks are distributed along the plateau region. Only those tunneling systems significantly contribute which have the right τ at a given activation frequency ω to obey the general relation $\omega\tau = 1$.

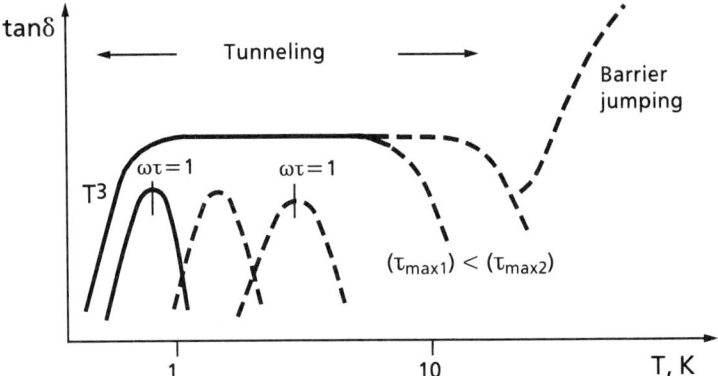

Fig. 6.14. Schematic plot of tan δ versus temperature at low frequencies in the plateau region.

The next selection process is governed by the transition rate which is the inverse relaxation time τ_{rel} given by Eq.(6.9). The transition rate between the two energy levels depends on the thermal energy k_BT compared to E. The case $E \ll k_BT$ is meaningless and $E \gg k_BT$ yields a negligible probability of transition. Qualitatively, an optimum exists for $E \approx 2k_BT$. This means that at each temperature T only those tunneling systems which have the proper value of E contribute significantly. Thus, the "effective" relaxation time is indirectly a function of the temperature. A broad range of E is reflected in a broad range of T (in the plateau region). Both, the values of E and τ are functions of the tunneling parameters Δ, Δ_0, $\Delta\phi$ and d (see Eqs.6.7 to 6.9). It can be shown that all tunneling systems involved have the same density if the tunneling parameters have a constant density distribution [6.8].

Despite the large reservoir of tunneling systems there are limitations. At very low temperatures no adequate tunneling systems can be activated anymore. The tunneling systems obeying the selection principles above die out, and tanδ drops. The temperature T_{min} which confines the plateau region depends on the applied frequency ω. This is demonstrated in Figure 6.15. At rather high frequencies (e.g., 15MHz) the drop starts at 1K; at low frequencies (e.g., 500Hz) a decrease is expected below 0.3K.

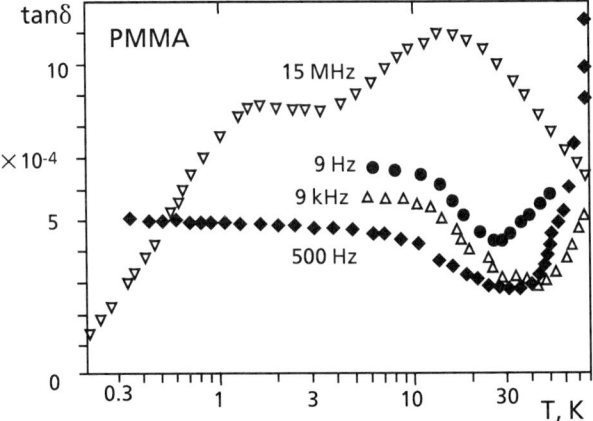

Fig. 6.15. Loss-factor tanδ for relaxation tunneling versus temperature at various phonon frequencies [6.11; 6.12].

Due to the large variety of relaxation times of tunneling systems no dependence on the measuring frequency is expected in the plateau region. This assumption has been verified in experiments, as the range of tan δ is not much changed when ω is varied by six decades (see Fig. 6.15).

According to Eqs.(6.7a) and (6.9), the smallest relaxation time τ_{min} is shortest for $\Delta=0$. This yields $E/\Delta_0=1$ in Eq.(6.9). τ_{min} is a reference value for the temperature dependence of the damping behavior below and at the plateau region. Assuming an experimental frequency ω, it can be shown that [6.8; 6.13]

$$\tan\delta \propto T^3/\omega \quad \text{for } \omega\tau_{min} \gg 1$$
$$\tan\delta \propto \text{const.} \quad \text{for } \omega\tau_{min} \ll 1.$$

The reservoir of contributing tunneling systems die out at low temperatures. The drop of tan δ with T^3 can be explained qualitatively by Eq.(6.9) when putting $E/\Delta_0=1$ and $E \propto k_B T$. The transition rate, which is proportional to τ^{-1}, controls the loss-factor.

The broadness of the plateau is indirectly determined by the limiting values of possible relaxation times. One confinement is τ_{min} at very low temperatures, which results from $\Delta=0$. The upper limit can be determined using Eq.(6.9):

$$\tau/\tau_{min} \approx (E/\Delta_0)^2 \tag{6.10}$$

For $\Delta_0 \to 0$ one gets $\tau \to \infty$. (For $\Delta_0=0$ there is no overlap of the wave functions of a double well potential and no more interaction). A finite value $\Delta_{0\,min}$ serves as cut-off parameter and defines τ_{max}. This parameter influences the reservoir of possible relaxation times. Calculations show that the plateau region spans a smaller temperature range if τ_{max} is small [6.11]. A schematic plot is shown in Figure 6.14. This might be an explanation of the strange behavior of PMMA and PBMA, mentioned in Section 6.4.1. For PMMA or PBMA there might be the case that the plateau region is too short and does not overlap sufficiently the jumping region. This leads to a damping minimum. The experimental low-temperature results of tan δ at low frequencies are shown in Figures 6.16 and 6.17.

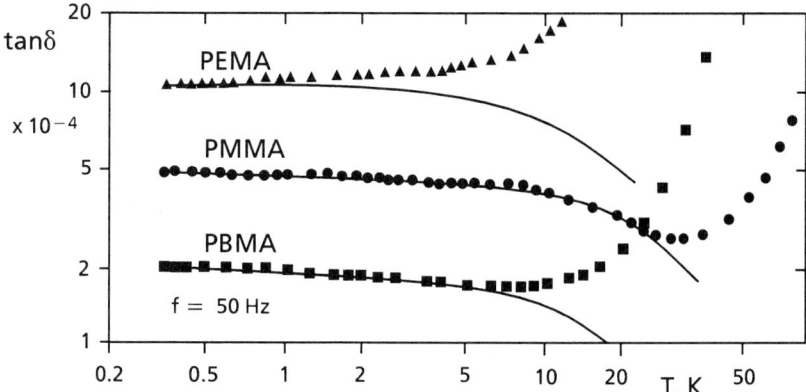

Fig. 6.16. Loss-factor versus temperature for PMMA, PEMA and PBMA; (frequency 50Hz).

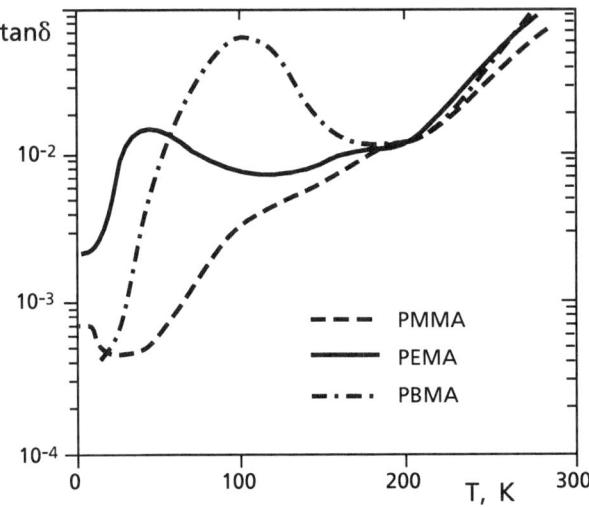

Fig. 6.17. Loss-factor versus temperature for PMMA, PEMA and PBMA (frequency: 9Hz).

Finally, a remark should be added on the range of relaxation times of tunneling systems. Under reasonable assumptions it can be shown [6.11]

$$\tau_{max} \approx 10^{12} \tau_{min} \tag{6.11}$$

This means that the relaxation times of tunneling systems are very widely spread and cover twelve orders of magnitude. (τ_{min}, at 1K is on the order of picoseconds).

6.6 ASYMPTOTIC BEHAVIOR AT LOW TEMPERATURES

Below 10K the loss-factor of amorphous polymers approaches a plateau region. For many amorphous polymers it has been found that the mechanical loss-factor at 4.2K (plateau region) lies in close vicinity of $\tan\delta \approx 10^{-3}$. This value is rather insensitive to the chemical composition, and seems to be a feature of tunneling processes; (jumping processes are no longer dominant at these temperatures). There are exceptions. For example, PEMA, PEI and PS show higher values, which are due to tails from the damping peaks around 40K. As already mentioned, PMMA and PBMA exhibit lower values and PMMA has a minimum around 20K. Most amorphous polymers, however, exhibit an asymptotic behavior which is shown in Figure 6.18. The solid curves characterize the limiting values of $\tan\delta$ found for the majority of amorphous polymers at low frequencies. It can be seen that $\tan\delta$ approaches a rather narrow band of values in the plateau region below 15K. As already mentioned, the plateau behavior results from tunneling processes. The plateau height is not strongly dependent on the frequency.

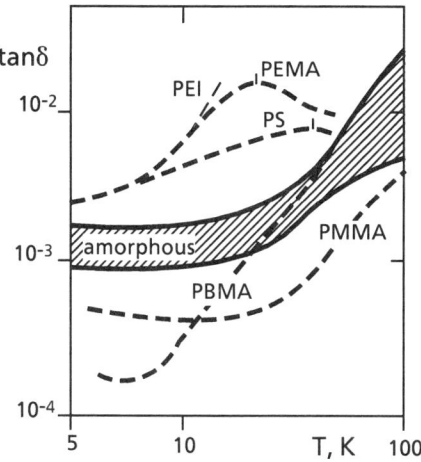

Fig. 6.18. Asymptotic damping behavior of amorphous polymers. The solid curves characterize the limiting values. The dashed curves show some exceptions.

Finally, a remark on semicrystalline polymers: The crystalline domains do not show secondary or tertiary glass transitions. The loss-factor of ideal crystallites is zero. Thus, the magnitude of the damping values is nearly proportional to the amorphous content. For semicrystalline polymers the value of $\tan\delta$ gets smaller when the amorphous phase is reduced. This is shown in Figure 7.15 for PTFE with different crystalline fractions f_c.

6.7 SUMMARY

Damping is mainly a feature of amorphous polymers. Dissipative processes with mechanical loading occur by place changes of molecular segments or side groups. Place changes occur by barrier jumping and tunneling processes.

Barrier Jumping

(1) Above 20K most place changes occur by jumping over potential barriers. The damping spectrum of secondary or tertiary glass transitions is characterized by damping peaks, which are specific of single molecular motions.

(2) The jumping processes are governed by thermal vibrations and take a characteristic time, the relaxation time. For single processes, the relaxation time is determined by the potential barrier height and the temperature. The Arrhenius equation establishes a relation between the relaxation time, the potential barrier and the temperature.

(3) A single relaxation time τ can be determined experimentally from a damping peak at a temperature T_m under cyclic loading with an angular frequency ω_m. At a damping peak: $\omega_m \tau(T_m) = 1$.

(4) At low temperatures the low-frequency damping values of most amorphous polymers approach asymptotically a narrow band ($\tan\delta \approx 10^{-3}$ at 4.2K), which is rather independent of the chemical composition.

Tunneling Processes

(1) Below 15K jumping processes are very unlikely and tunneling processes dominate which are responsible for the plateau region. Only place changes of small molecular units are involved.

(2) Tunneling processes in terms of quantum-mechanics can be characterized by a transition between two-energy levels, which are given by the height and thickness of potential barriers and the energy difference of potential minima of a double well potential.
(3) For relaxation tunneling processes incoming phonons modulate the energy levels, thus activating transitions between the two-energy level systems. Resonant tunneling processes occur by absorption of phonons (or photons), which have the energy of the level difference E.
(4) The plateau region at very low temperatures is confined by a steep drop of the damping values; $\tan\delta \propto T^3/\omega$.
(5) The relaxation times of tunneling systems span a range of many decades.

6.8 REFERENCES

6.1 Hunklinger,S.and A.K.Raychandhur; Z. Phys. B., Condensed Matter 57, (1984) and Progr. in Low Temp. Phys. IX, (1986) p. 265; Ed. D.F. Brewer, Elsevier Sci Publ.

6.2 Hunklinger, S.; in : "Phonon Scattering in Condensed Matter ," Solid State Science 51, (1984) 378, Eds.: W. Eisenmenger, K. Laßmann and S. Döttinger, Springer Verlag, Berlin.

6.3 Mc Crum, N.G. in Ref. 1.2, p. 10.

6.4 Schatzki, T.K.; J. Polymer Sci. 57 (1962) 496.

6.5 Shimizu, K.; O.Yano and Y.Wada; J. Polymer Sci. 13 (1975) 2357 (Polymer Phys. Ed.).

6.6 Koppelmann, J.; Progr. Colloid + Polymer Sci. 66 (1979) 235.

6.7 Heijboer,J.and M.Pineri;in: "Nonmetallic Materials and Composites at Low Temperatures", Vol. II (1980), p. 89, Eds. Hartwig, G.; Evans, D.;-Plenum Press; New York.

6.8 Shimizu, K.; O.Yano, and Y.Wada; J. Polymer Sci. 11 (1973) 1644 (Polymer Phys. Ed.).

6.9 Federle,G. and S.Hunklinger; in: "Nonmetallic Materials and Composites at Low Temperatures," Vol. II (1980), p. 49, Eds. Hartwig, G.and Evans, D.;-Plenum Press.

6.10 Geiß, N.; G.Kaspar and S.Hunklinger; in: "Nonmetallic Materials and Composites at Low Temperatures," Vol. III (1986), p. 99, Eds. Hartwig, G. and Evans, D.; Plenum Press.

6.11 Hickel, W. and G.Kaspar; Cryogenics 28 (1988).

6.12 Hartwig, G.; in Ref. 1.8.

6.13 Ahlborn, K.; Cryogenics 28 (1988) 234.

6.14 Blochinzew, D.J.; Grundlagen der Quantenmechanik (1953), p. 350, Deutscher Verlag der Wissenschaften, Berlin.

6.15 from Ref. 1.4 p. 166.

ERRATA

The correct key to Figure 2.20 (page 41) should be as shown here:

        ▦ longitudinal,
        ☐ transverse,
        ▨ optical.

Page 138 was omitted from Chapter 6. Please see the reverse of this sheet for that page.

Polymer Properties at Room and Cryogenic Temperatures   0-306-44987-0
Günther Hartwig   Plenum Press, New York, 1994

MECHANICAL DEFORMATION BEHAVIOR

Contents

7	Mechanical deformation behavior	141
7.1	General survey	141
7.2	Definition of moduli and compliances	142
7.3	Deformation characteristics of polymers	143
7.4	Constant load step	144
7.5	Cyclic loading	151
7.6	Dependence on crystallinity	155
7.7	Constant load rate	157
7.8	Deformation at high load levels	161
7.9	Asymptotic behavior at low temperatures	165
7.10	Poisson's ratio	168
7.11	Oriented polymers	169
7.12	Summary	170
7.13	References	171

Appendix

7A : Tables

7 MECHANICAL DEFORMATION BEHAVIOR

7.1 GENERAL SURVEY

The correlation between applied mechanical load and resulting deformation is described by moduli. The basic component of moduli is controlled by binding forces, originating from deformations of the electron configurations. The load acts against the binding forces. This component reacts nearly immediately and is responsible for the elastic behavior. At very low temperatures most polymers behave elastically. Below 20K, the moduli of amorphous, semicrystalline and cross-linked polymers, respectively are rather similar which suggests similar binding forces. Above 30K the asymmetry of binding potentials causes a small decrease of the moduli owing to thermal expansion of the chain distances. The response, however, is strictly elastic. In the vicinity of a glass transition temperature (dispersion region) viscoelastic processes decrease the moduli owing to unfreezing of molecular motions. Only secondary or tertiary transitions are considered here. The deformation behavior in a dispersion region depends on temperature and time. The typical deformation characteristic is plotted schematically for the Young's modulus E in Figure 7.1 with three characteristic regions:

(a) constant elastic modulus (symmetric potential),
(b) temperature-dependent elastic modulus (asymmetric potential),
(c) time- and temperature-dependent viscoelastic modulus near a glass transition temperature T_g. The latter is time-dependent, for example, a function of the strain rate $\dot{\varepsilon}$ or the frequency ω.

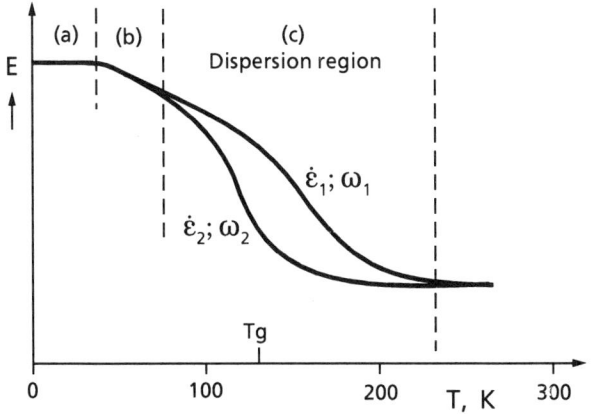

Fig. 7.1. Young's modulus E versus temperature.
$\dot{\varepsilon}$: strain rate
ω: frequency
$\dot{\varepsilon}_1 > \dot{\varepsilon}_2$; $\omega_1 > \omega_2$

Because of their time dependence, the moduli are a function of the time profile of loading. A definition of special moduli is made for each loading type. The most important loading profiles and their defined moduli are:

- Step function with a constant load → relaxation modulus.
- Constant load rate → initial modulus, secant modulus.
- Cyclic loading → complex modulus.

By means of the superposition principle it is possible to correlate the deformation response between different load-time functions. If viscoelastic processes are involved, strain- or stress-controlled loading yields a somewhat different result, and a distinction between modulus and compliance is made. Only in the elastic case is the modulus equal to the reciprocal compliance. Below 150K and at low loads, a distinction between strain- and stress-controlled loading is negligible. The most important moduli (or compliances) for different load types are summarized in Table 7.1.

Another distinction of deformation characteristics is given by thermodynamics. Each deformation is related to a minimum of the free energy $F = U - ST$.

- In the **energy elastic** case, a mechanical load changes only the internal energy U. After load release the original energy is restored by thermal vibrations, which are the driving forces.
- In the **entropy elastic** case, the load only decreases the entropy S by molecular orientations. After load release the isotropic state will be restored; the driving force is the entropy. Entropy elasticity is dominant mainly in elastomers at elevated temperatures, and is not the topic of this book.

7.2 DEFINITION OF MODULI AND COMPLIANCES

A modulus E or compliance F is defined for uniaxial strain- or stress-controlled loading, respectively.

$$\sigma = E\,\varepsilon \quad \text{or} \quad \varepsilon = F\,\sigma \qquad (7.1)$$

Analogous parameters are defined for shear load. As already mentioned, different moduli are introduced for loads with specific time profiles (see Table 7.1). These moduli are time- and temperature-dependent at elevated temperatures. The relaxation- or creep moduli depend on the elapsed time t of constant load, the complex moduli on angular frequency ω, the initial moduli on the strain rate $\dot{\varepsilon}$, and the secant moduli are functions of the strain rate $\dot{\varepsilon}$ and the maximum strain ε_m applied. The analogous relations hold for the compliances. At low temperatures it is not important to distinguish between relaxation modulus and creep compliance; the universal term relaxation is used.

The moduli, except the secant modulus, are defined at low loads. They are valid in the region of linear viscoelasticity, where the moduli do not depend on the load amplitude. The secant modulus, however, is of more general nature and depends on strain- or stress amplitudes. It is used for describing nonlinear viscoelastic (or viscous) deformations.

MECHANICAL DEFORMATION BEHAVIOR

Table 7.1 Moduli and compliances.

Moduli and compliances	uniaxial	shear	load-time profile
Relaxation modulus	$E(t,T)$	$G(t,T)$	constant load step
Relaxation compliance	$F(t,T)$	$J(t,T)$	
Complex modulus	$E^*(\omega,T)$	$G^*(\omega,T)$	cyclic loading
Complex compliance	$F^*(\omega,T)$	$J^*(\omega,T)$	
Dynamic modulus [1]	$E'(\omega,T)$	$G'(\omega,T)$	cyclic loading
Dynamic compliance	$F'(\omega,T)$	$J'(\omega,T)$	
Damping modulus	$E''(\omega,T)$	$G''(\omega,T)$	cyclic loading
Damping compliance	$F''(\omega,T)$	$J''(\omega,T)$	
Initial modulus	$E(\dot{\varepsilon}, T)$	$G(\dot{\gamma}, T)$	constant load rate
Initial compliance	$F(\dot{\sigma}, T)$	$J(\dot{\sigma}_\tau, T)$	
Secant modulus	$E(\dot{\varepsilon}, \varepsilon_m, T)$	$G(\dot{\gamma}, \gamma_m, T)$	constant load rate
Secant compliance	$F(\dot{\sigma}, \sigma_m, T)$	$J(\dot{\sigma}_\tau, \sigma_{\tau m}, T)$	

Nomenclature

Load	Modulus	Compliance	Stress	Strain
uniaxial	E	F	σ	ε
shear	G	J	σ_τ	γ

Superposition Principle. The deformation behavior for loads with different time profiles can be correlated by the superposition principle. If one knows the time dependence of a modulus, e.g., $E(t)$ and the time profile of strain $\varepsilon(t)$, then the time dependence of the stress $\sigma(t)$ can be calculated. An example will be given:

$$\sigma(t) = E_0 \varepsilon(t) + \int_0^t \dot{E}(t-x) \, \varepsilon(x) dx = \int_0^t E(t-x) \, \dot{\varepsilon} \, dx \qquad (7.2)$$

with $\dot{E}(t-x) = \dfrac{dE(t-x)}{d(t-x)}$ and E_0 the elastic (fast reacting) modulus [7.1].

The time variable x accounts for the different starting times of relaxation processes during loading. An important assumption of the superposition principle is that relaxations are not influenced by preloadings. This is true for viscoelastic behavior without irreversible components. An example for applying the superposition principle will be demonstrated in Figure 7.18.

7.3 DEFORMATION CHARACTERISTICS OF POLYMERS

External forces cause deformations of molecular segments and of electron orbitals in polymers. The following deformation types are fundamental:

---
[1] Is sometimes called storage modulus.

Elastic response, originating from deformation of electron orbitals owing to binding forces (fast components). The relaxation time is about 10^{-13}s [7.2]. For polymers mainly weak Van der Waals binding forces between chains are involved which are responsible for the low moduli of elasticity. (The deformation of the molecular chains is negligible in isotropic polymers.) This type of deformation is

> reversible, nondissipative, time- and nearly temperature- independent.

(the anharmonicity of binding potentials causes a small temperature dependence). A model for describing deformations in terms of the chain arrangements is given in Figure 7.2. Elastic deformations are sketched on rigid chain segments at different arrangements: (a) entangled (b) aligned perpendicular and (c) parallel to the load direction.

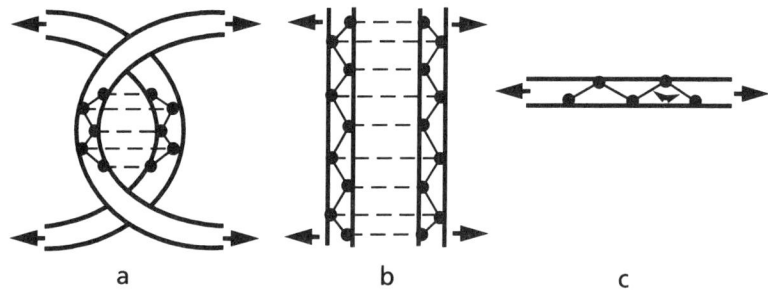

Fig.7.2. Elastic deformation of rigid chain segments; (dashed lines symbolize the Van der Waals forces). **a)** entangled, **b)** aligned perpendicular to the load direction, **c)** aligned parallel to the load direction.

Between the entangled chain segments the repulsive component of Van der Waals forces controls the deformation behavior (Fig. 7.2a). For an array perpendicular to the force, the attractive Van der Waals forces react (Fig. 7.2b). A deformation of chain components can be neglected because of their strong covalent binding forces. The modulus can be calculated from the binding potentials (second derivative with respect to the chain distance; see Eq.(7.25)).

Viscoelastic response, resulting from a time-consuming reorientation of molecular segments or side groups by barrier jumping (slow component). Only neighboring potential minima are involved. (Their description has already been given in Section 6.3). This type of deformation is

> reversible, dissipative, time- and temperature-dependent.

Linear viscoelastic response exists at low loads or at very low temperatures; otherwise a nonlinear viscoelastic response prevails.

Viscous response, resulting from molecular slipping. This occurs when the loading energy plus thermal energy of molecules is high enough to surpass several potential barriers in the Van der Waals potential distribution. This type of deformation is

> nonreversible, dissipative, time- and temperature-dependent.

Below 100K and for not too high loads, this deformation is negligible. At higher load levels, however, irreversible deformations occur even at a temperature as low as 77K.

At elevated temperatures, chain segments are no longer rigid and viscoelastic or viscous deformations reduce the effect of binding forces. These deformation characteristics can be described by models, one of which is shown in Figure 7.3. It comprises a time-independent spring F_0, which represents the fast, pure elastic components (binding forces). The combination of a spring F_1 and a time-dependent dashpot η_1 represents the slow, reversible viscoelastic component. A dashpot η_2 accounts for irreversible processes.

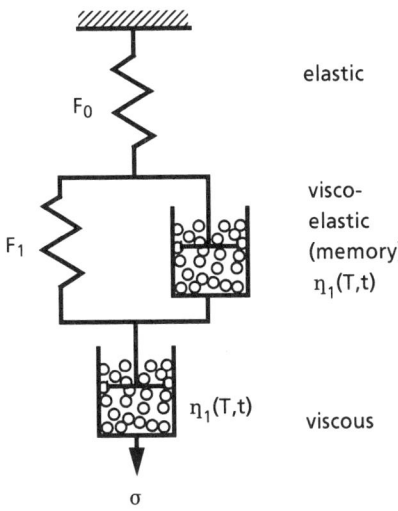

Fig. 7.3. Model showing elastic, viscoelastic and viscous components.

This model describes single relaxation processes, which prevail at secondary or tertiary glass transitions. (In the general case of different molecular mobilities an array of springs and dashpots is applied). At low temperatures, the dashpots become more and more frozen and only elastic components remain. Especially the viscous component is small or negligible below 100K and low loads.

7.4 CONSTANT LOAD STEP

7.4.1 General Description

When applying a load step only the elastic components react immediately, and a slow creep- or relaxation behavior from viscoelastic processes follows. Asymptotically a saturation is achieved which corresponds to the relaxed state under load. When suddenly releasing the load, similar processes restore the original situation (irreversible processes are negligible at low temperatures and with moderate loads). The principle is shown in Figure 7.4.

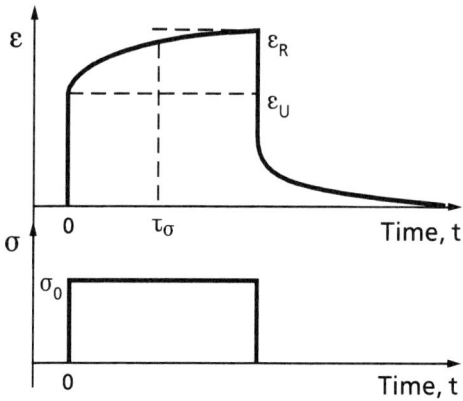

Fig. 7.4. Creep behavior at a stress step.

This behavior can be described by a time-dependent modulus or compliance. The compliance at a stress step σ_0 (uniaxial load) or $\sigma_{\tau 0}$ (shear load) is defined as:

$$F(t,T) = \varepsilon(t,T)/\sigma_0 \qquad (7.3a)$$

$$J(t,T) = \gamma(t,T)/\sigma_{\tau 0} \qquad (7.3b)$$

and analogously for the modulus at a strain step ε_0 (uniaxial) or γ_0 (shear):

$$E(t,T) = \sigma(t,T)/\varepsilon_0 \qquad (7.4a)$$

$$G(t,T) = \sigma_\tau(t,T)/\gamma_0 \qquad (7.4b)$$

These parameters can be calculated from the relaxation times, which are specific for viscoelastic processes and the load type applied. An example is given in Figure 7.5.

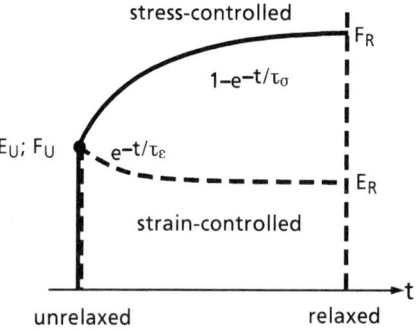

Fig. 7.5. Time-dependent modulus E(t,T) and compliance F(t,T) at a constant strain- or stress step, respectively.

E_u and F_u are the unrelaxed modulus and compliance, respectively. $E_R(T)$ and $F_R(T)$ are the asymptotically relaxed values under load at a temperature T.

τ_ε and τ_σ are the temperature-dependent relaxation times at constant strain and stress, respectively. For example, τ_ε is the time at which (E_u-E_R) is reduced by the factor 1/e. For a single relaxation process and uniaxial load the relations hold:

$$E(t,T) = E_R(T)+(E_u-E_R(T))e^{-t/\tau_\varepsilon} \qquad (7.5a)$$

$$F(t,T) = F_u+(F_R(T)-F_u)(1-e^{-t/\tau_\sigma}). \qquad (7.5b)$$

If several relaxation processes with τ_i contribute, a summation has to be performed involving a relaxation time spectrum E_i or a retardation time spectrum F_i:

$$E(t,T) = E_R(T) + \Sigma_i E_i e^{-t/\tau_{ei}} \qquad (7.5c)$$

$$F(t,T) = F_u + \Sigma_i F_i (1 - e^{-t/\tau_{oi}}) \qquad (7.5d)$$

Analogous relations hold for shear loading.

7.4.2 Unrelaxed State

The question arises what is the meaning of an unrelaxed state of a polymer? At an infinitesimal steep initial load step no time is available for any relaxation and one always will get the same initial value E_u at any temperature T. This is the real unrelaxed state of a polymer. At $t>0$, relaxation processes start which, of course, depend on temperature. At elevated temperatures this state exists only for a very short time when load is applied. The situation is shown schematically in Figure 7.6 by solid lines.

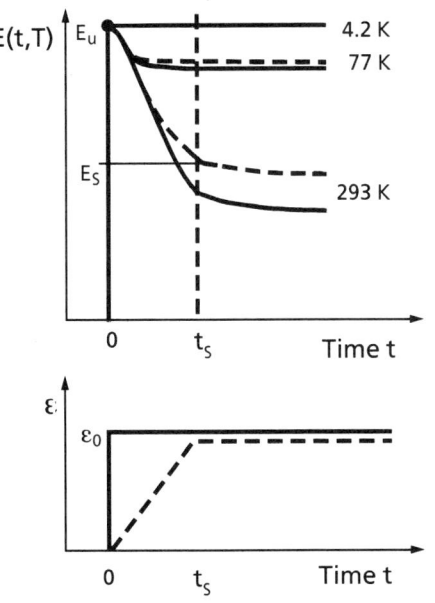

Fig.7.6. Relaxation modulus $E(t,T)$ at strain-controlled loading.

———— very steep strain step
- - - - ramp

The unrelaxed modulus is equal to the modulus measured at very low temperatures, e.g., at 4.2K. Very little relaxation occurs, and E is nearly time-independent. At RT, however, many relaxation processes are activated which reduce the modulus drastically.

Under real conditions an infinitesimal steep load step is impossible and a certain time t_s (called loading time) is necessary for applying a load. Within this time period already several relaxations can occur. When varying the loading time t_s, the apparent modulus can be varied. This is a consequence of the time dependence of viscoelastic processes. If a load ramp is applied, fewer relaxation processes are activated at the time t_s than for an infinitesimal steep step. The dashed lines in Figure 7.6 indicate the situation of a ramp. In a real experiment t_s is of the order of 0.1s. One finds a modulus E_s, which is called the starting modulus. It is the initial value for measuring creep or relaxation as a function of time. Therefore generally, E_s, and F_s should be used instead of E_u and F_u in Eqs. (7.5a,b). The fast relaxations already occur during initial loading. What is measured in a relaxation experiment are the remaining slow components. (The analogous situation exists at cyclic loading with low and very high frequencies; see Section 7.5).

7.4.3 Results

The time dependence of the compliance F at constant stress is plotted in the Figures 7.7 to 7.9 (The curves can be described by Eq.(7.5d) which involves several retardation processes). Since the initial load step is not steep ($t_s \approx 0.1$s), the value of the starting compliance F_s has been used at the beginning of the creep measurements. Since t_s usually is very short, it can be neglected for quoting the total load duration. The decisive parameter is the relative change between F_s and the asymptotically relaxed compliance F_R at a sufficient load duration. In Figure 7.7 the behavior at 77K is shown for several polymers. The polymer POM shows nearly no relaxation and HDPE a relatively large one. In Figures 7.8 and 7.9 the time dependence of F at various temperatures is shown for HDPE and PTFE. In the vicinity of the secondary glass transition temperature (≈ 155K for HDPE and ≈ 173K for PTFE) the relative change of F is of the order of 30 % to 40%. HDPE exhibits some creep behavior ($\approx 2\%$) even at 4.2K. Cross-linked polymers (epoxy resins) show a relatively weak relaxation behavior, even at elevated temperatures (see Table 7.2).

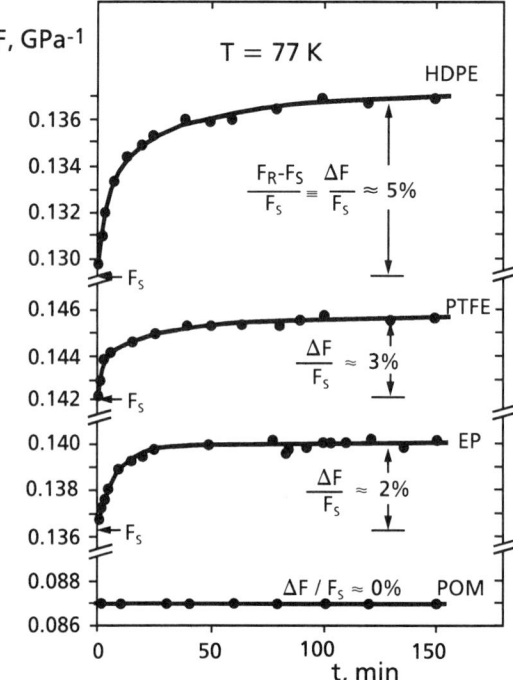

Fig. 7.7. Relaxation compliance F(t,T) at 77K versus the creep time

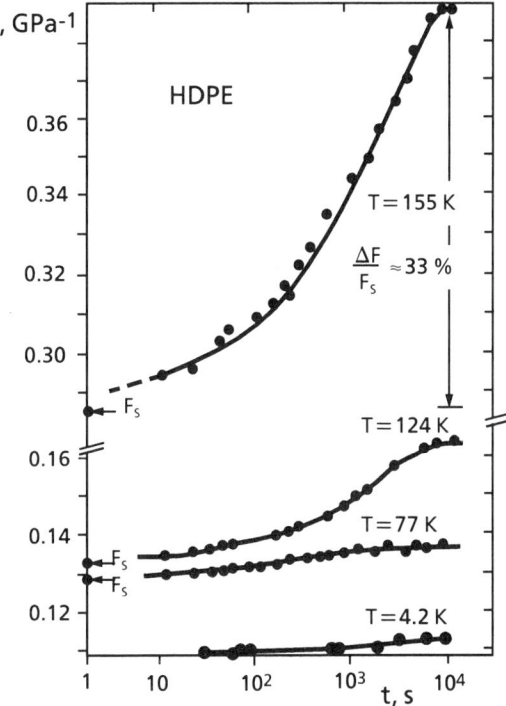

Fig. 7.8. Relaxation compliance F(t,T) of HDPE at various temperatures versus creep time.

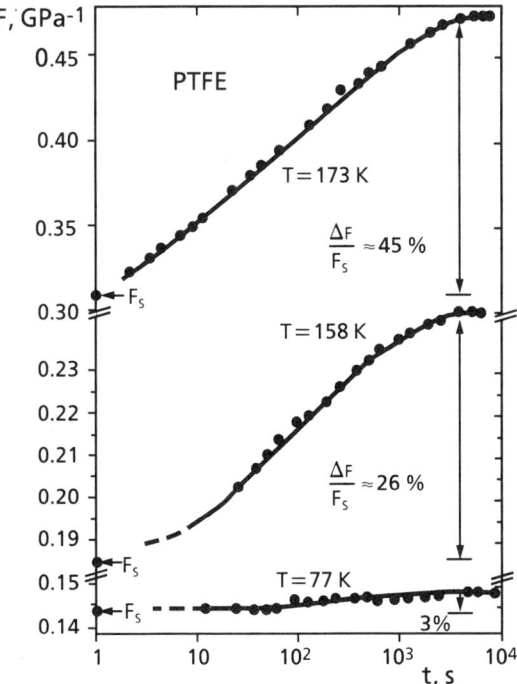

Fig. 7.9. Relaxation compliance F(t,T) of PTFE at various temperatures versus creep time.

The relative change $\Delta F/F_s$ is compiled for several polymers at various temperatures in Table 7.2.

Table 7.2. Relative change of the compliance.

Polymer	T, K	$\Delta F/F_s$, %	T, K	$\Delta F/F_s$, %
PC	187	17	—	—
HDPE	150	30	77	5
PTFE	150	25	77	3
EP	150	12	77	2

When plotting the curves of Figures 7.7 to 7.9 in a double-logarithmic scale, straight lines are found which obey the relations:

$$\frac{d\ln F}{d\ln t} = m_F(T) \qquad (7.6a)$$

$$\frac{d\ln E}{d\ln t} = -m_E(T) \qquad (7.6b)$$

At low temperatures, stress- and strain-controlled measurements yield nearly the same result for modulus E and compliance F; thus $E \approx 1/F$. Below 150K: $m_F(T) \approx m_E(T) = m(T)$. In Table 7.3 values of the relaxation parameter m(T) are compiled.

Table 7.3. Relaxation parameters m (T).

Polymer	T	m (T)
HDPE	T = 158K	0.043
	T = 124K	0.02
	T = 77K	0.006
	T = 4.2K	0.002
PTFE	T = 173K	0.053
	T = 158K	0.032
	T = 77K	0.003
EP	T = 151K	0.017
	T = 77K	0.003
PC	T = 187K	0.018
	T = 77K	0.005

7.5 CYCLIC LOADING

7.5.1 General Description

The following description assumes low load amplitudes and a quasi-linear viscoelastic behavior. For sinewave loading a phase shift δ occurs between stress and strain when viscoelastic damping processes are involved. This is shown in a vector diagram in Figure 7.10.

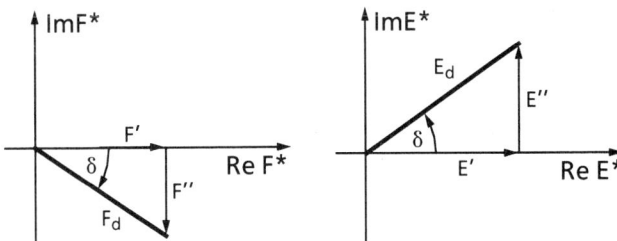

Fig. 7.10. Vector diagram for modulus E and compliance F.

A complex representation is useful for cyclic loading:

$$E^*(\omega,T) = E'(\omega,T) + i\,E''(\omega,T) \equiv E_d(\omega,T)\,e^{i\delta} \qquad (7.7)$$

$$F^*(\omega,T) = F'(\omega,T) - i\,F''(\omega,T) \equiv F_d(\omega,T)\,e^{-i\delta} \qquad (7.8)$$

E' and F' are the dynamic (or storage) modulus and dynamic compliance, respectively. They are not identical to the elastic modulus or compliance which describe only the fast components controlled by the binding forces. They involve frequency- and temperature-dependent place changes. Basically, they are analogous to the starting modulus E_s or compliance F_s defined in Figure 7.6. Only as T→0, or ω→∞ are place changes frozen, and the unrelaxed modulus E_u (or compliance F_u) remains.

E" and F" are the damping components.

E_d and F_d are the absolute values of the modulus- or compliance vector, which can be determined from the maximum stress σ_0 and maximum strain ε_0 for cyclic loading. The following relation holds for the modulus

$$E_d(\omega,T) = \left(E'^{\,2} + E''^{\,2}\right)^{1/2} = \sigma_0/\varepsilon_0 \qquad (7.9)$$

At low temperatures E" is negligible and E' ≈ E_d. Below 200K is E"/E' < 0.1.

Generally, moduli and compliances are correlated by their complex values

$$E^* F^* = 1 \qquad (7.10)$$

The relation E' = 1/F' is exact only for the elastic case without viscoelastic contributions. However, at low temperatures this relation is a good approximation in the viscoelastic case. Stress- or strain-controlled cyclic loadings yield nearly the same results. Analogous relations apply to shear loading.

7.5.2 Correlations of Moduli and Damping

The relaxation moduli, the complex moduli and the damping behavior are all correlated. Their frequency- and temperature dependences can be calculated by using these correlations. The dependence on angular frequency ω can be calculated if the unrelaxed modulus E_U and the relaxed modulus $E_R(T)$ are known. For a single relaxation[2] process with a relaxation time $\tau_\varepsilon(T)$, the following general relations hold [7.3]

$$E'(\omega,T) = E_R(T) + \left(E_U - E_R(T)\right) \frac{\omega^2 \tau_\varepsilon^2}{1+\omega^2 \tau_\varepsilon^2} \quad (7.11)$$

$$E''(\omega,T) = \left(E_U - E_R(T)\right) \frac{\omega \tau_\varepsilon}{1+\omega^2 \tau_\varepsilon^2} \quad (7.12)$$

This is true for strain-controlled measurements. For stress-controlled measurements the relaxation processes have a somewhat higher relaxation time $\tau_\sigma(T)$. It can be shown [7.3]:

$$\frac{\tau_\varepsilon}{\tau_\sigma} = \frac{E_R}{E_U} \quad (7.13)$$

At low temperatures (T<100K) $\tau_\varepsilon \approx \tau_\sigma$ and in the range of strong secondary glass transitions: $\tau_\varepsilon/\tau_\sigma \approx 0.6$ to 0.8.

It follows from Eq.(7.12) that a maximum of the damping modulus E" occurs at $\omega\tau_\varepsilon = 1$, and analogously for the damping compliance F" at $\omega\tau_\sigma = 1$. The frequency dependences of E', F' and E", F" are plotted schematically in Figure 7.11. A sligthly different situation is true for the commonly used damping quantity, the loss-factor tanδ which is defined by

$$\tan\delta = \frac{E''}{E'} = \frac{F''}{F'} \quad (7.14)$$

This ratio exhibits a damping maximum at a somewhat different frequency-temperature position than E"or F". From Eqs. (7.11) and (7.12) one gets

$$\tan\delta = \frac{(E_U - E_R)\,\omega\tau_\varepsilon}{E_R + E_U \omega^2 \tau_\varepsilon^2} \quad (7.15a)$$

If one defines $\tau = (E_U/E_R)^{1/2} \tau_\varepsilon$, the well-known relation follows

$$\tan\delta = \frac{E_U - E_R}{(E_U E_R)^{1/2}} \frac{\omega\tau}{1+\omega^2\tau^2}. \quad (7.15b)$$

It further can be shown that $\tau = (\tau_\varepsilon \cdot \tau_\sigma)^{1/2}$. A maximum of tanδ exists for $\omega\tau = 1$. The frequency- or temperature dependence of deformation and damping is compiled in Figure 7.11.

---
[2] If different relaxation processes contribute a summation must be performed which is weighted by a relaxation time spectrum.

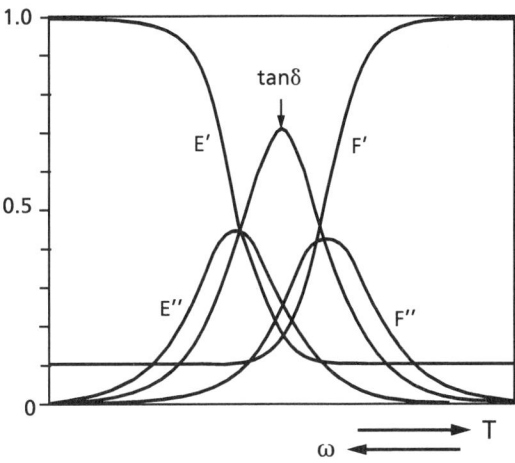

Fig. 7.11. Schematic frequency- and temperature dependence of modulus E', compliance F' and the damping components E", F" and tanδ.

It can be seen that maxima of E", F" and tanδ occur at different frequencies or temperatures due to different relaxation times. In the region of strong secondary glass transitions $\tau/\tau_\varepsilon \approx 1.2$ to 1.4. Below 100K or at very high frequencies it nearly holds: $\tau_\varepsilon \approx \tau_\sigma \approx \tau$ and the maxima of E", F" and tanδ coincide.

Each relaxation time is a function of temperature and can be calculated by the Arrhenius equation (6.3). The temperature shift of the damping maxima E" and tan δ resulting from different values of τ_ε and τ is small. For example, at $\omega = 36$Hz the maxima of E" and tanδ for HDPE occur at 154K and 158K, respectively. For maxima at higher temperatures the differences become larger. For further treatment no distinction is made between the different kinds of relaxation times; $\tau \equiv \tau_\varepsilon \approx \tau_\sigma$ is used as an approximation. Both the modulus step and the damping maxima can be shifted by temperature- or frequency variations as shown in Figure 7.12.

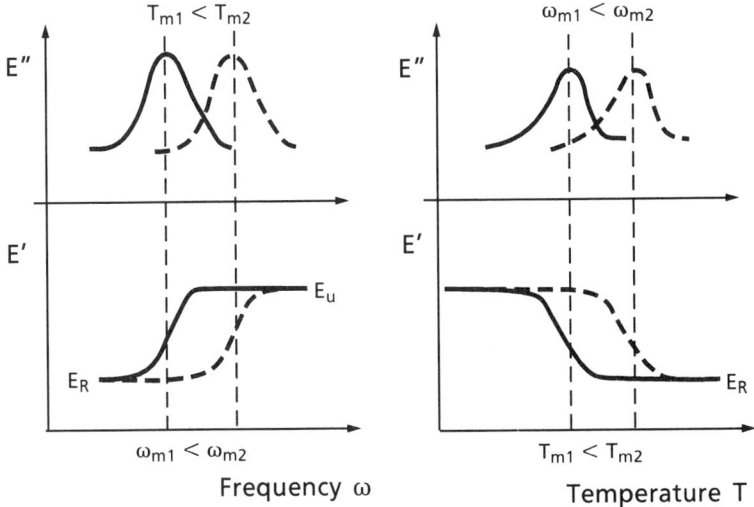

Fig.7.12. Frequency-and temperature shifts of the damping- and storage components of a modulus.

For $\omega\tau \ll 1$, the relaxation time is small enough that molecular motions follow the activation immediately. For this case the modulus corresponds to the relaxed modulus E_R, which is no longer time-dependent.

For $\omega\tau \gg 1$, the relaxation time is too long to follow the short activation time and no relaxation can be activated. The modulus then corresponds to the unrelaxed modulus E_U. Only if the relaxation time is of the order of the activation time, exactly at $\omega\tau = 1$, does a change of modulus occur. If different relaxations are involved several modulus steps are superimposed.

7.5.4 Results and Discussion

The storage moduli E' for tensile loading and G' for shear loading are plotted versus temperature in Figures 7.13 and 7.14. The modulus E' has been measured at different frequencies which, in the dispersion region, causes a modulus shift. For comparison the loss-factor tanδ is shown as an indicator of a relaxation process.

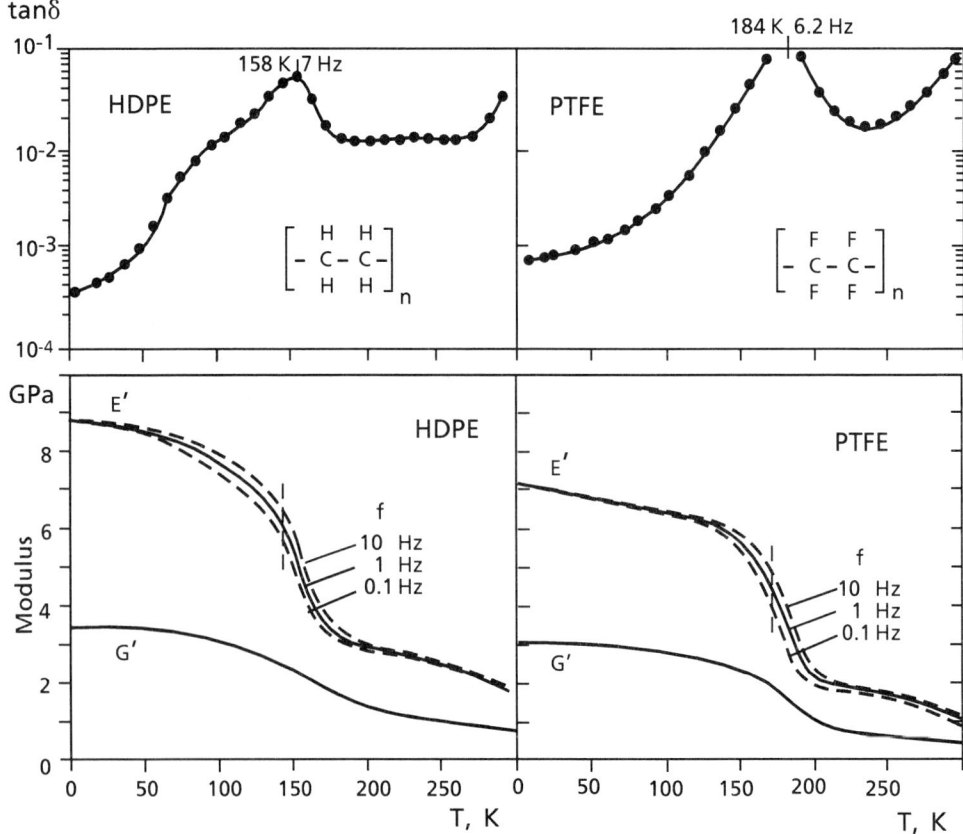

Fig. 7.13. Dynamic moduli and tanδ versus temperature. E' is measured at different frequencies f, and G' at 5Hz.

In the dispersion region the temperature dependence is strong whereas the influence of frequency is less pronounced. A frequency change by a factor of 100 has less effect than a 20 K temperature change. In the dispersion region the modulus curves are shifted by less than 20 K when the frequency is varied from 0.1 Hz to 10 Hz.

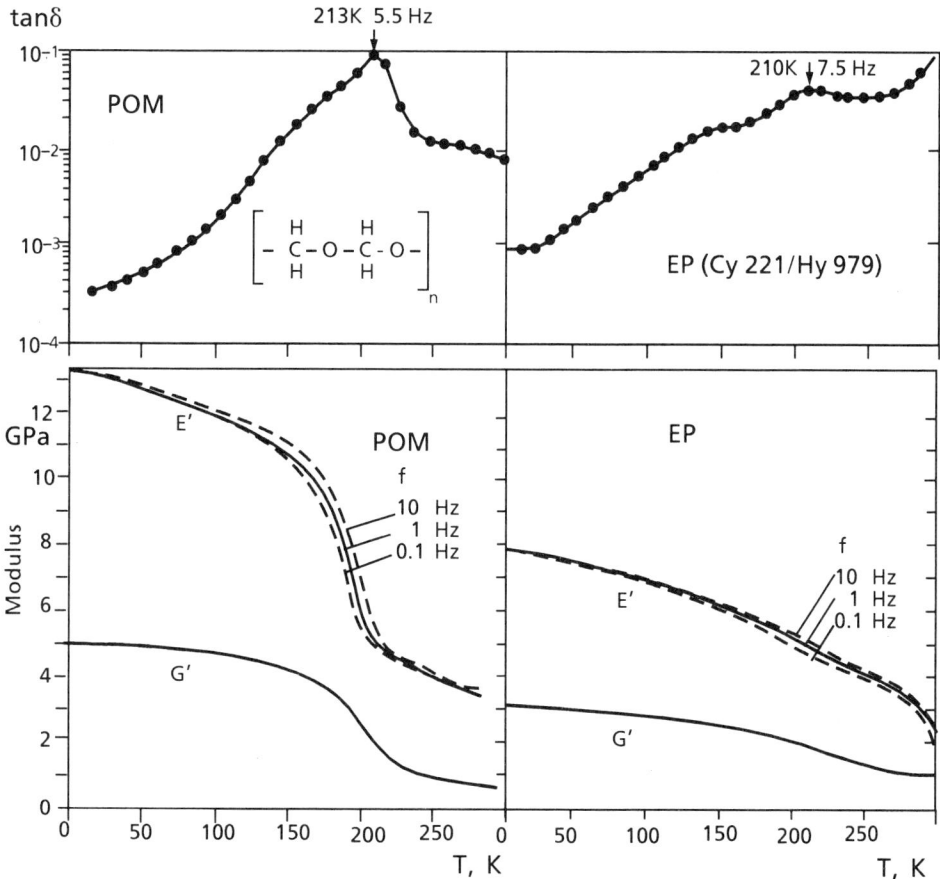

Fig. 7.14. Dynamic moduli and tan δ versus temperature. E' is measured at different frequencies f; G' only at 5 Hz (Epoxy resin: Cy 221 / Hy 979; Ciba Geigy).

7.6 DEPENDENCE ON CRYSTALLINITY

At very low temperatures no large difference exists between the modulus of amorphous and crystalline polymers. In the vicinity of secondary or tertiary glass transitions, however, viscoelastic processes decrease the modulus of the amorphous domains. This is demonstrated in Figure 7.15 for PTFE with different crystalline contents [7.4]. For a high crystalline content, the shear modulus G' stays nearly constant and only low damping components exist. A more amorphous modification of PTFE shows a larger loss-factor and modulus step.

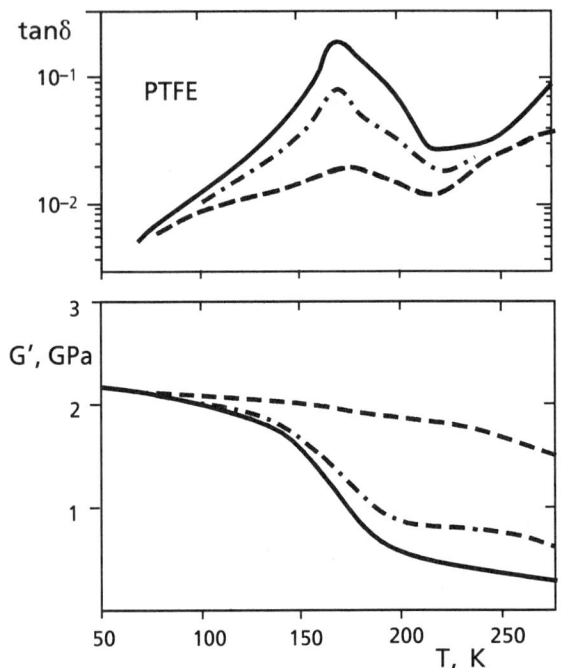

Fig. 7.15. Loss-factor tan δ and shear modulus G' of PTFE with different crystalline contents f_c [7.4].

- - - - $f_c = 92$ vol %
- · - $f_c = 60$ vol %
———— $f_c = 42$ vol %

The modulus of a semicrystalline polymer is determined by the moduli E'_C and E'_A of the crystalline and amorphous phases, respectively. A model of Takayanaki takes account of both serial and parallel connections of E'_C and E'_A relative to the load direction. A schematic picture is shown in Figure 7.16.

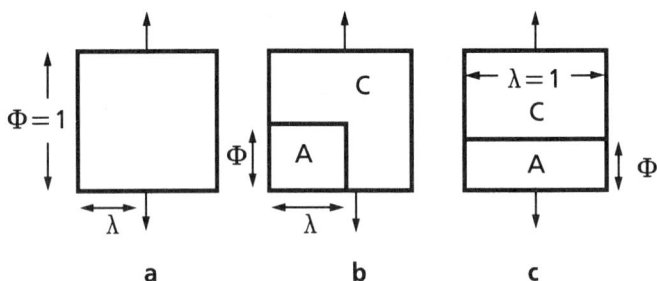

Fig. 7.16. Takayanaki model: **a)** parallel array; **b)** serial and parallel array; **c)** serial array of the amorphous (A) and crystalline domains (C).

From Figure 7.16b the general relation is easily verified [7.5]

$$E'_{semi} = \left[\frac{\Phi}{\lambda E'_A + (1-\lambda) E'_C} + \frac{1-\Phi}{E'_C} \right]^{-1} \tag{7.16}$$

The large modulus steps of semicrystalline polymers at secondary glass transitions can be explained by a serial connection of E'_C and E'_A according to Figure 7.16c. This is consistent with the results shown in the Figures 7.13 and 7.14.

7.7 CONSTANT LOAD RATE
7.7.1 General Description and Results

Deformations at a constant load rate exist in many technical applications. Three types of moduli can be deduced from stress-strain diagrams:

- Initial modulus (obtained from the slope at $\varepsilon \to 0$)
- Secant modulus
- Tangent modulus.

The initial modulus describes only the linear behavior at low loads. The secant modulus more generally is used for describing the nonlinear behavior. (The tangent modulus will not be treated here). It will be started with a short survey of the stress-strain behavior.

The stress - strain behavior generally shows several characteristic regions:
- **Quasi-linear viscoelastic** region, realized at small load amplitudes or at low temperatures.
- **Nonlinear viscoelastic** region, occuring at medium load amplitudes and at elevated temperatures.
- **Nonlinear viscoelastic - viscous** region, occuring at nearly ultimate load conditions well below RT.
- **Yielding and necking** region, arising from irreversible shear flow processes at nearly ultimate load conditions and at high temperatures.

In the **quasi-linear viscoelastic** region stress-strain diagrams are reversible; the diagrams are nearly similar for loading and unloading; hysteresis effects are small and correlated to the loss-factor tanδ. Strain- or stress-controlled loadings yield very nearly the same stress-strain diagrams.

In the **nonlinear viscoelastic** region the stress - strain diagrams are not quite reversible. Small strains of the order of 0.2% to 0.5% remain even at 77K immediately after unloading. After about 1 hour, however, some of the remaining strain relaxes. Thus, only small irreversible deformation occurs. Characteristic nonlinear behavior is more pronounced for thermoplastic polymers. Cross-linked polymers (e.g., epoxies) are rather linear and reversible below 100K (see Fig. 9.16).

In the **nonlinear viscoelastic-viscous** region, loading and unloading yield different stress-strain curves. Hysteresis effects are much larger than determined from the loss-factor tanδ. They become larger still at increased load amplitudes. This is shown in Figure7.24. Load levels in the vicinity of fracture stress cause large deformations, but no yielding occurs below about 100K. An important irreversible deformation process is the craze formation [7.6]. Several polymers, especially thermoplastics, tend to this process at elevated temperatures. Yielding and necking only occur at elevated temperatures. It is a prestep to fracture and will be treated in Section 9.3.4.

7.7.2 Initial Modulus

The initial modulus describes the linear viscoelastic components of a deformation. It is determined from the initial slope at low loads. In Figure 7.17a several stress-strain diagrams are plotted schematically with different strain rates $\dot{\varepsilon}$ and at different temperatures. At 4.2K, nearly no $\dot{\varepsilon}$ dependence exists and the diagrams are rather linear. At 77K some influence of $\dot{\varepsilon}$ exists. The initial slope increases with the strain rate and, at an extremely large $\dot{\varepsilon}$, one would get a slope, which corresponds to an unrelaxed modulus as discussed in the previous section. In the vicinity of a glass transition temperature T_g, the $\dot{\varepsilon}$ dependence is pronounced and hysteresis effects become visible. Experimental curves at 77K on PSU are shown in Figure 7.17b (Similar curves will be shown in Figure 9.10 for PC). The shape of the stress-strain diagrams depends on the load rate. The initial slope is a function of $\dot{\varepsilon}$ but it is only slightly changed by a large variation of $\dot{\varepsilon}$. The slope changes by 15% when $\dot{\varepsilon}$ is varied by a factor of 7000. The dependence on $\dot{\varepsilon}$ would be more pronounced at a secondary glass transition. For epoxy resins, the strain rate dependence is generally smaller than for thermoplastic polymer [7.7].

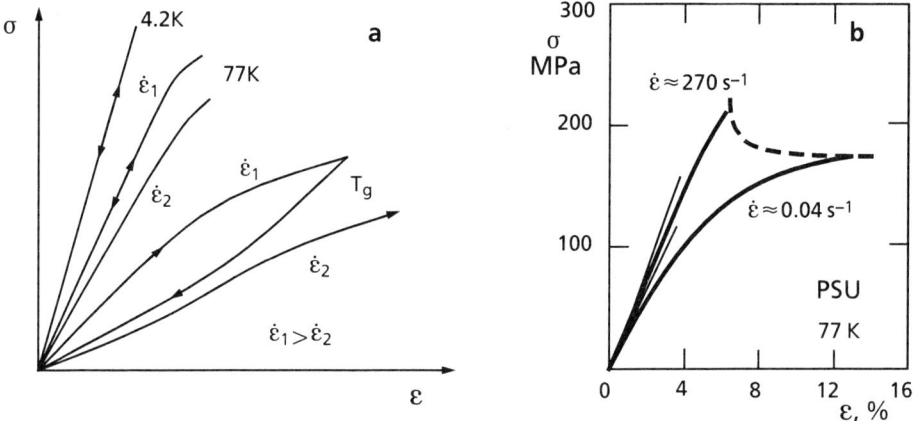

Fig. 7.17. a) Schematic stress-strain diagrams at different strain rates $\dot{\varepsilon}$ and different temperatures. b) Experimental stress-strain diagrams for PSU at 77K at different strain rates $\dot{\varepsilon}$. (The dashed line is the fracture line; see Section 9.4)

7.7.3 Secant Modulus

If viscoelastic or viscous processes are involved, a stress-strain diagram becomes nonlinear at higher loads. The explanation of the nonlinear stress-strain behavior which determines the secant modulus can be seen from Figure 7.18. At each infinitesimal strain step there are activated time-dependent relaxations, which decrease the stress $\sigma(t)$ and the relaxation modulus $E(t)$ (see Fig. 7.5). Successive strain steps are more and more influenced by the relaxations of the preceeding steps, and a nonlinear behavior results. The shape of the stress-strain diagrams can be calculated by the superposition principle.

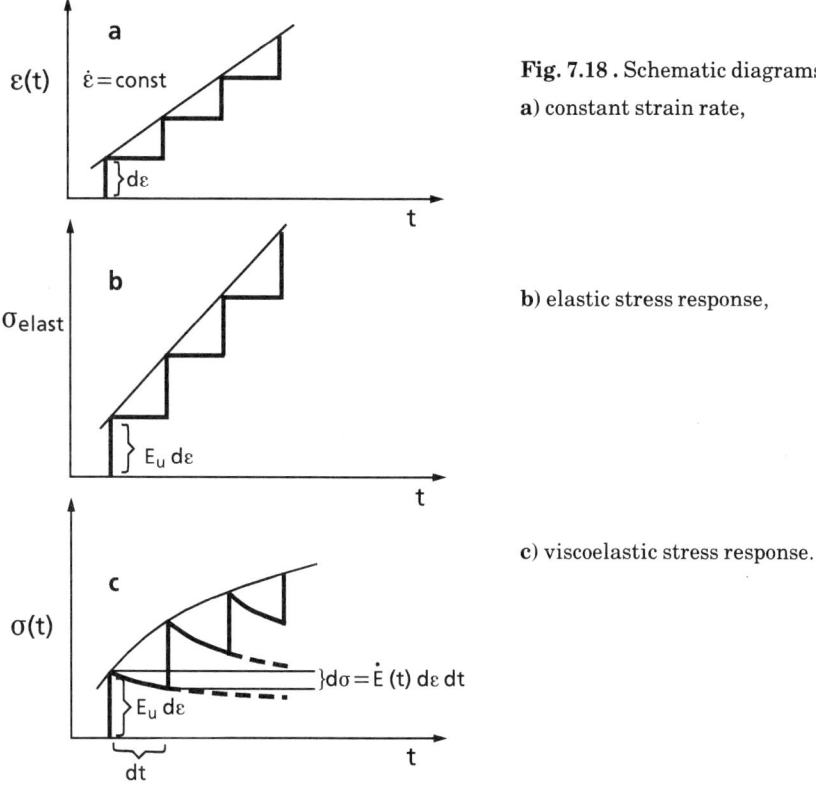

Fig. 7.18. Schematic diagrams
a) constant strain rate,

b) elastic stress response,

c) viscoelastic stress response.

The secant modulus can be used for describing the nonlinear behavior. A schematic diagram is shown in Figure 7.19.

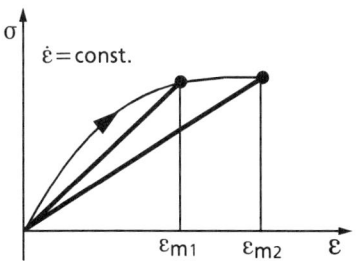

Fig. 7.19. Schematic stress-strain diagram at constant strain rate $\dot{\varepsilon}$ and different maximum strains ε_{m1} and ε_{m2}.

In the nonlinear viscoelastic region, the secant modulus depends in addition on the load amplitude. During the loading time more and more relaxations are activated. At a constant $\dot{\varepsilon}$, the loading time t_m increases with the maximum strain ε_m applied. This is indicated by ε_{m1} and ε_{m2} in Figure 7.19. The loading time t_m is determined from

$$t_m = \varepsilon_m / \dot{\varepsilon} \tag{7.17}$$

The secant modulus E is defined as

$$E(\dot{\varepsilon}, \varepsilon_m, T) = \sigma(t_m, T)/\varepsilon_m \quad (7.18)$$

The measured stress $\sigma(t,T)$ is assumed to depend on time and temperature. For the secant modulus the stress at the time $t=t_m$ is the decisive parameter, which can be calculated by the superposition principle. Typical results of the secant modulus are given in Figure 7.20 as a function of $\dot{\varepsilon}$. The modulus has been determined at two different maximum strains $\varepsilon_{m1} < \varepsilon_{m2}$. As expected from Figure 7.19, the values of E are smaller for ε_{m2} since more time for relaxations is available.

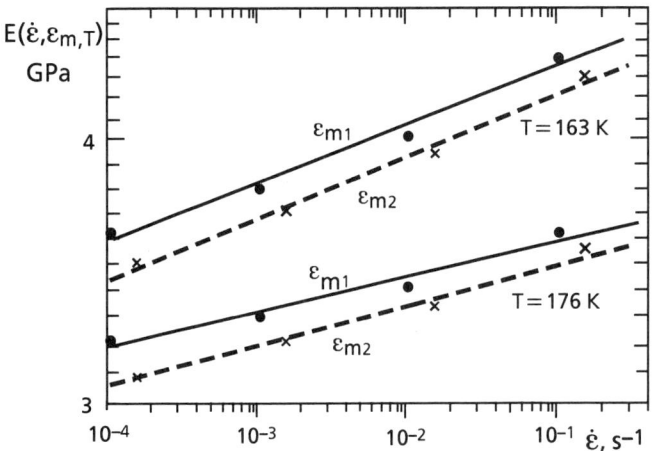

Fig. 7.20. The secant modulus E of HDPE versus strain rate $\dot{\varepsilon}$ at two different temperatures and two maximum strains $\varepsilon_{m1} = 1\%$ and $\varepsilon_{m2} = 1.6\%$.

7.7.4 Discussion

It can be seen from Figure 7.20 that the slope of E(t,T) depends on temperature and is constant in a double-logarithmic plot.

$$\frac{d \ln E}{d \ln \varepsilon} = \text{const.}(T) \quad (7.19)$$

This relation is similar to Eq.(7.6b) and can be used for calculating the modulus as a function of $\dot{\varepsilon}$. A basic deduction, however, is given by the superposition principle which will be used for calculating $\sigma(t_m, T)$ after the loading time t_m. From Eq.(7.2) one gets

$$\sigma(t_m, T) = \int_0^{t_m} E(t_m - t)\, \dot{\varepsilon}\, dt \quad (7.20)$$

The time dependence of the relaxation modulus E can be described by the relaxation parameter m(T) defined by Eq.(7.6b). After some calculation one gets the final result for the secant modulus [7.8]

$$E(\dot{\varepsilon}, \varepsilon_m, T) \approx E_s(T) \left(1 + m(T) \ln \frac{\dot{\varepsilon} t_g}{\varepsilon_m} \right) \tag{7.21}$$

$E_s(T)$ is the starting modulus and t_g a reference time which can be eliminated when comparing the moduli at different strain rates $\dot{\varepsilon}_1, \dot{\varepsilon}_2$ and strains $\varepsilon_{m1}, \varepsilon_{m2}$.

$$E(\dot{\varepsilon}_2, \varepsilon_{m2}, T) = E(\dot{\varepsilon}_1, \varepsilon_{m1}, T) + E_s(T) \left\{ \ln \frac{\dot{\varepsilon}_2}{\dot{\varepsilon}_1} - \ln \frac{\varepsilon_{m2}}{\varepsilon_{m1}} \right\} m(T) \tag{7.22}$$

This equation can also be used for calculating the initial moduli at different strain rates, when putting $\varepsilon_{m1} = \varepsilon_{m2} \to 0$ in Eq. (7.22).

Results on the strain rate dependence of stress-strain diagrams are shown in Figures 7.17b and 9.12c. The initial slopes of stress-strain diagrams for PSU in Figure 7.17b are changed by 15%, when the strain rate is varied by a factor of 7000. A similar result has been found for PC in Figure 9.12c. In this case $\dot{\varepsilon}$ is varied by a factor of 17000 and the initial slopes are changed by 10%. This is in rough agreement with Eq.(7.22), when taking the value of m=0.005 at 77K from Table 7.3. Only a small modulus change is achieved by a very large variation of the strain rate. In the dispersion region, however, the increase of the modulus is larger since the relaxation parameter m is higher (see Table 7.3).

7.8 DEFORMATIONS AT HIGH LOAD LEVELS

At high load levels and not too low temperatures the deformation behavior is nonlinear and not quite reversible. Several deformation processes are involved and the nonlinear behavior of polymers arises from a material specific mixture of processes. Only some important processes or tendencies will be considered. Typical stress-strain diagrams at different temperatures are plotted schematically in Figure 7.21. Experimental curves are shown in Figures 7.22a,b at 4.2K and 77K. As expected, the diagrams at 4.2K are rather linear up to fracture and nearly independent of the load rates. At 77K, however, only cross-linked polymers such as EP exhibit a rather elastic behavior. Thermoplastic polymers, by contrast, are rather nonlinear and PE shows a stress maximum which usually is attributed to a yield point. But no necking occurs at 77K. Most polymers at low temperatures break prior to surpassing a yield point.

At high temperatures (or in the region of a strong glass transition) two characteristic phenomena occur after the yield point: strain softening and strain hardening (see Fig.7.21). The first effect can be explained by a higher flow rate due to a shorter load assisted relaxation time, the second one by orientations.

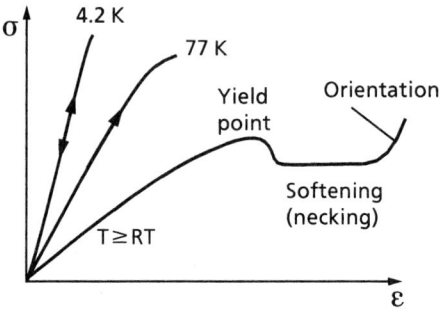

Fig. 7.21. Schematic stress-strain curves at different temperatures and load rates.

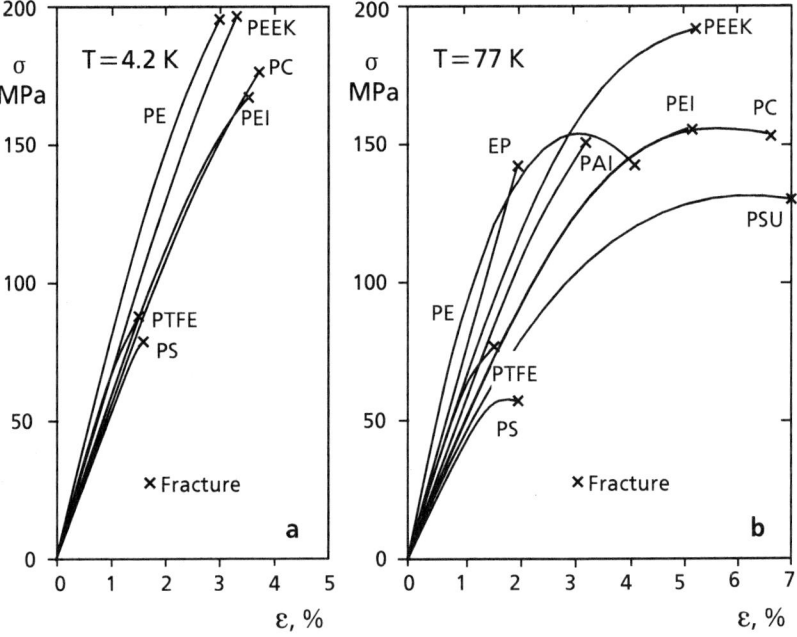

Fig. 7.22. Stress-strain curves of several polymers at 4.2K and 77K

Load Assisted Relaxation Time. Apparently the binding forces are reduced at high loads. Several investigations revealed that no increase of the total free volume is induced by loading, but it is deformed in the load direction [7.9]. A modified Arrhenius equation has been proposed by Eyring. The mean potential barrier $\Delta\phi$ between two minima is decreased by the mean external work $\Delta w(\sigma)$, which depends on the external load amplitude σ. (The situation with and without external work w is plotted in Figure 6.10). The relaxation time is thus a function of T and σ. The modified Arrhenius equation yields the load assisted relaxation time

$$\tau_i(T,\sigma) = v_0^{-1} \exp\left(\frac{\Delta\phi - \Delta w(\sigma)}{RT}\right) = v_0^{-1} \exp\left(\frac{\Delta\phi_i - g_i \sigma}{RT}\right) \tag{7.23}$$

where $v_0 \approx 10^{13}$Hz is the thermal frequency; in a rather rough approximation $\Delta w \approx g_i \sigma$; and g_i is a material parameter correlated to the free volume available for a segmental mobility i. The relaxation time gets shorter with increased loads. A feedback is established between the relaxation time and the load amplitude. Due to a shorter relaxation time, the jumping rate is increased and nonlinear effects are enhanced. This effect becomes dominant in the region of strain softening. (It is, however, realistic to assume that some softening effects already occur prior to the yield point by load assisted relaxation times). The influence of the stress σ on the relaxation time τ_i can be measured on creep experiments at different load levels. The relaxation times from Eq.7.23 are used in Eqs.7.5a,b. At a constant temperature usually several relaxation processes contribute during the creep time. A summation has to be performed according to Eqs.7.5c,d.

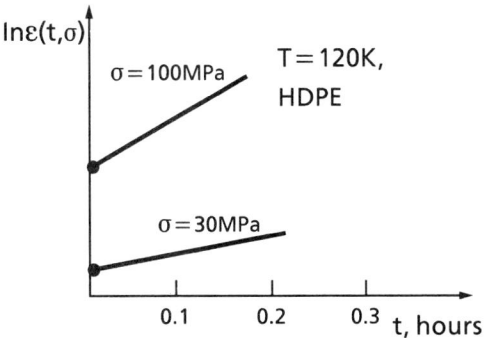

Fig. 7.23. Strain relaxation at 120K versus time for different initial stress steps.

Experimental results at 120K are shown in Figure 7.23 for HDPE. Relaxation curves during low and high load steps exhibit a different behavior. The dependence on time of the strain relaxation at a low stress step (linear viscoelastic region) differs from that at a high step (viscoelastic-viscous region). The relaxation time τ_σ is shorter at a high stress, which is evident from the steeper slope in the semilogarithmic plot in Figure 7.23. The slope is proportional to the inverse relaxation time τ_σ (see Eqs.7.5a,b). It can be assumed that at 120K during the first minutes of creep time mainly the strong glass transition at 155K of HDPE contributes. The curves can be described approximately by a single relaxation process. Additional information can be obtained from successive reloading tests.

7.8.2 Loading, Unloading and Repeated Loading

The shape of stress-strain curves is changed after each load cycle and depends on the load amplitude of the previous load cycle. Successive loading and unloading curves at different stress levels are shown in Figure 7.24. At low load levels, loading and unloading is reversible and connected with only small hysteresis. At high load levels, hysteresis is large and a small strain remains after unloading (it relaxes, however, after some time, about 1 h at 77K). The explanation results from the shorter load assisted relaxation time at increased load (Eq.(7.23).) There is a difference in the relaxation behavior induced by loading and by unloading. Place changes occur at a higher rate during loading and at a somewhat lower rate for unloading. (This is, of course, only true if the loading time is of the same order as the relaxation time).

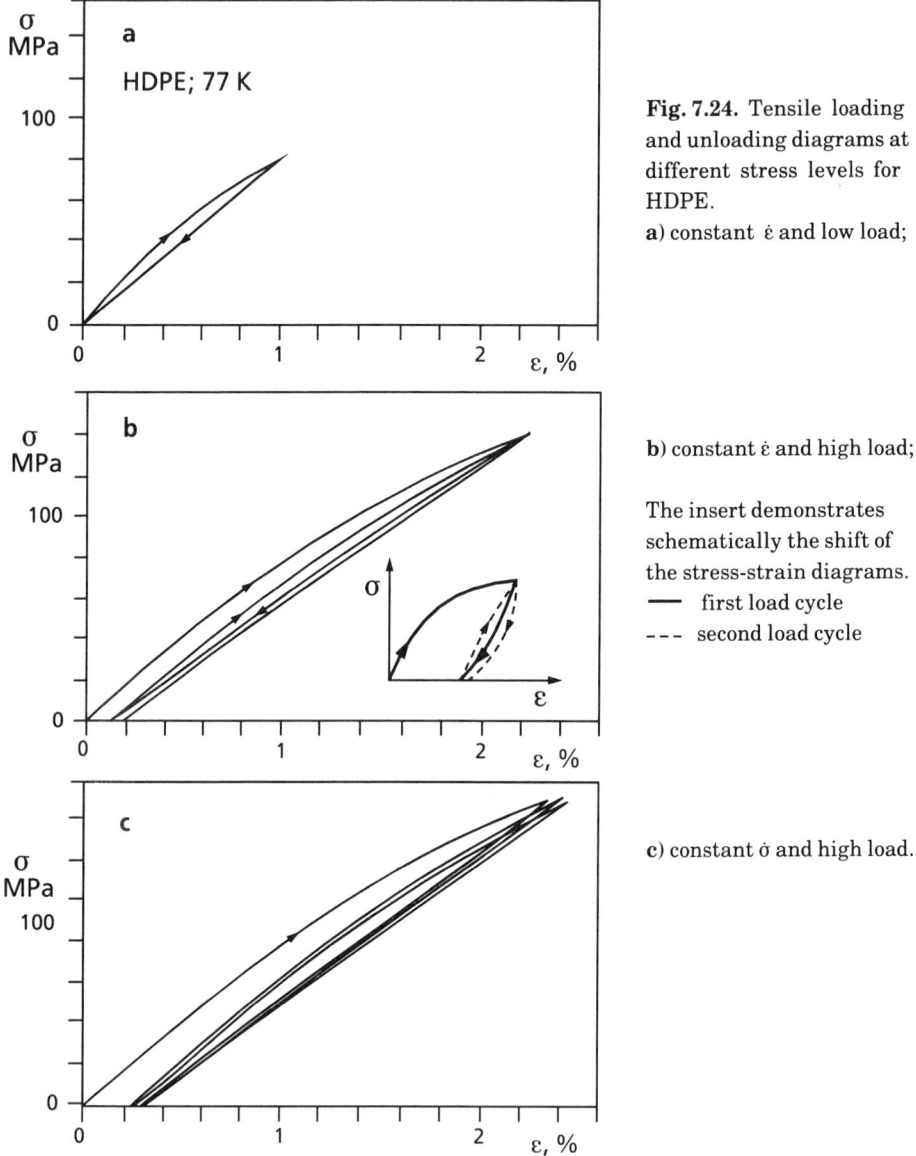

Fig. 7.24. Tensile loading and unloading diagrams at different stress levels for HDPE.
a) constant $\dot{\varepsilon}$ and low load;

b) constant $\dot{\varepsilon}$ and high load;

The insert demonstrates schematically the shift of the stress-strain diagrams.
— first load cycle
--- second load cycle

c) constant $\dot{\sigma}$ and high load.

During loading the relaxation time gets shorter and the rate of place changes is more and more increased. The situation is reversed for unloading. The relaxation time becomes longer at decreased load and the rearrangement of place changes is not yet finished after unloading. Immediate reloading starts therefore from a state of nonequilibrium. Further immediate load cycles exhibit less relaxations and a more linear behavior than the first cycle. The hysteresis effects become smaller and stabilize at a certain value after some cycles. Most of the energy is dissipated during the first cycle and less in the following ones. This result is important for fatigue cycling since it controls the heating of specimens. The diagrams are little different for strain- and stress-controlled loading, as seen from Figures 7.24b and 7.24c. But again, the tendency is obvious that hysteresis effects are decreased in successive load cycles. The situation is slightly different for alternative tension-compression loading. In this case, after unloading of tensile stress (or strain), compressive forces are immediately applied which restore the remaining stress- or strain components and induce nonlinear deformations by compression. The hysteresis areas are smaller and are again reduced upon further cycling (see Fig. 7.25). The considerations of this section are important for fatigue cycling, which will be treated in Section 9.5.

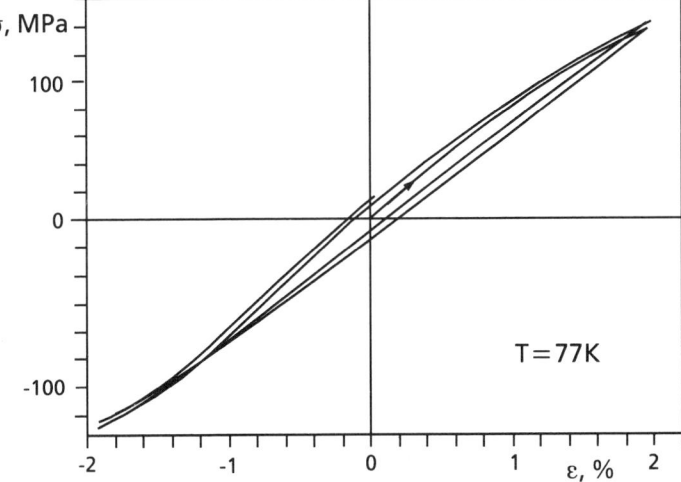

Fig. 7.25. Stress-strain diagrams for alternating loading in tension and compression (at a constant strain rate).

7.9 ASYMPTOTIC BEHAVIOR AT LOW TEMPERATURES

7.9.1 Moduli at Very Low Temperatures

At very low temperatures and not too high stress levels the deformation behavior is determined mainly by the Van der Waals binding potential. Time-dependent viscoelastic effects are negligible. Experiments reveal that even the chemical compositions of polymers have no large influence on the low-temperature deformation behavior. Cross-linking and the crystalline fraction, however, are parameters, which determine to some extent the moduli. Highly crystalline or cross-linked polymers tend to higher moduli than those of amorphous thermoplastic polymers. The dynamic- or the initial moduli (they are equal at very low temperatures) will be considered in this section. In Figures 7.26 and 7.27 the values of elasticity moduli E' and shear moduli G' in the low-temperature region are compiled for several polymers. It can be seen that at very low temperatures the moduli are nearly independent of temperature and that they asymptotically approach narrow bands of values each for the amorphous and the semicrystalline polymers. Epoxy resins with different chemical compositions and cross-link densities also approach similar values. The experimental data suggest that at very low temperatures the moduli of polymers are not very sensitive to the chemical compositions. They depend, however, on morphological parameters such as crystalline fraction or cross-linking. The ranges of values are slightly higher for semicrystalline and cross-linked polymers than for thermoplastic polymers.

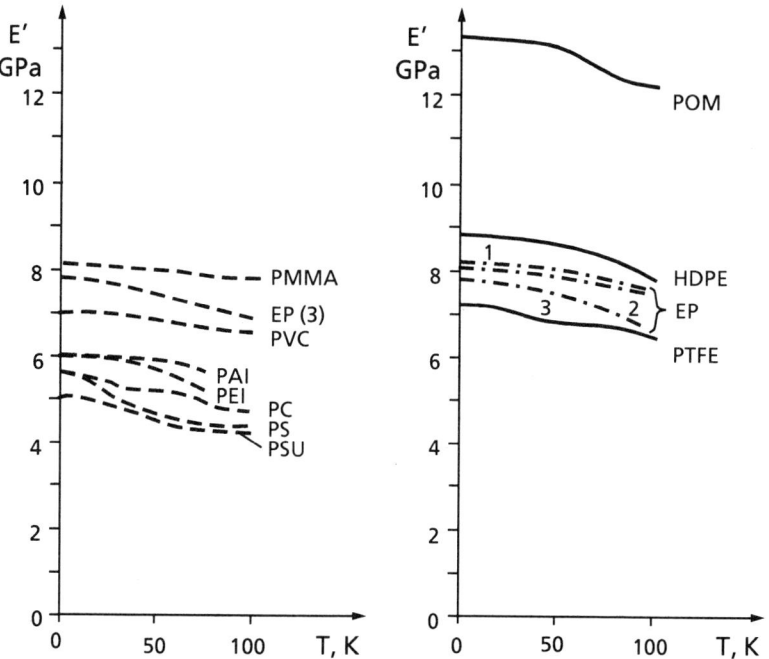

Fig.7.26. Young's modulus E' versus temperature. —— semicrystalline; – – amorphous; – · – cross-linked. Epoxies: (1) X183/HY 905; (2) MY740/Jeffamin D230; (3) CY221/HY 979; (Ciba-Geigy)

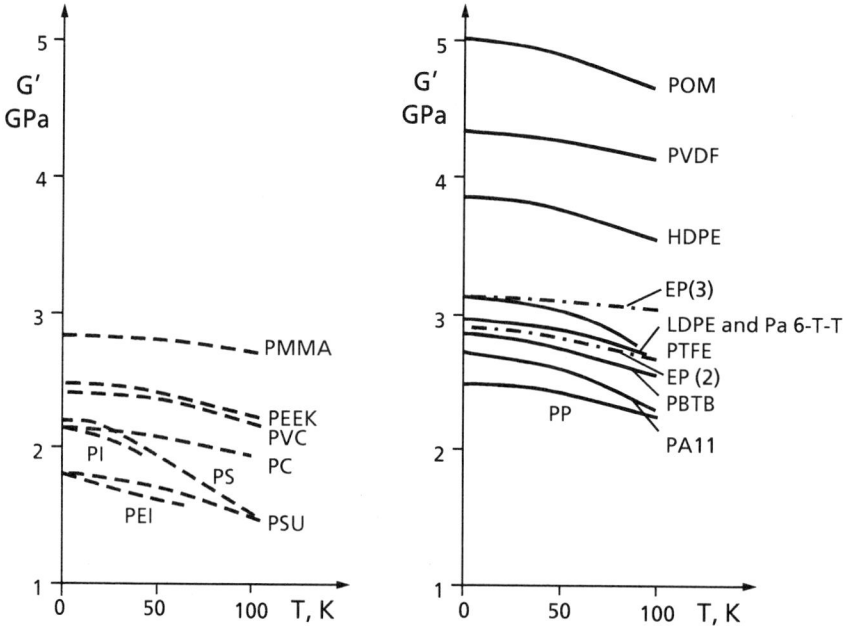

Fig.7.27. Shear modulus G' versus temperature. —— semicrystalline; – – – amorphous; – · – cross-linked.

MECHANICAL DEFORMATION BEHAVIOR

The values of moduli at T→0 are within the following ranges:

- semicrystalline and   E' = 7 - 9 GPa   G' = 2.5 - 4 GPa
  cross-linked polymers

- amorphous polymers   E' = 5 - 8 GPa   G' = 1.8 - 3 GPa

An exception is POM which shows excessively high moduli. (More data are compiled in Appendix 7A)

7.9.2 Moduli and Binding Potentials

The moduli are determined by the fast elastic components of the modulus which are controlled by the Van der Waals binding forces. Thus, the binding forces are similar for amorphous and for semicrystalline polymers, respectively. This can be seen from the correlation of the binding potential ϕ and the modulus E. The binding force P is given by

$$P = d\phi_w / dr_w \qquad (7.24a)$$

where dr_w is the mean distance change of chains in the Van der Waals potential ϕ_w (intrachain forces can be neglected). The stress is defined by the force P per area A

$$\sigma = P / A \qquad (7.24b)$$

The area A is given by $A \approx r_c r_w$, where r_c and r_w are the atomic- and chain distances, respectively. By taking $d\varepsilon = dr_w/r_w$, the modulus can be calculated from

$$E' = d\sigma/d\varepsilon = \frac{d^2\phi_w}{dr_w^2} \frac{r_w}{A} \approx \frac{d^2\phi_w}{dr_w^2} \frac{1}{r_c} \qquad (7.25)$$

The second derivative of the potential is proportional to its curvature. Since these considerations apply to very low temperatures, the following conclusions can be drawn for the potential minima.[3] The curvature of the Van der Waals potential minimum is similar for most polymers because the moduli are similar. For semicrystalline or cross-linked polymers the curvature is slightly more pronounced than for amorphous thermoplastic ones.

---

[3] These considerations are not limited to the potential minima, which are symmetric and yield temperature independent moduli. An asymmetric potential distribution results from $r_c = r_c(T)$, due to thermal expansion.

7.10 POISSON'S RATIO

For uniaxial loading the cross section perpendicular to the load direction is changed. For strain loading the following definition holds for the Poisson's ratio

$$\mu' = -\varepsilon_\perp/\varepsilon_\parallel \tag{7.26}$$

where $\varepsilon_\perp$ is the transverse strain induced by a longitudinal strain $\varepsilon_\parallel$.
Analogous to the complex moduli, the Poisson's ratio can be represented as a complex value

$$\mu^* = \mu' - i\mu'' \tag{7.27}$$

μ'', however, is usually very small and can be neglected. For isotropic materials the values of μ' are between 0.2 and 0.5; the latter value applies to rubber. The Poisson's ratio is a measure of the free volume available. At glass transitions the Poisson's ratio is influenced by viscoelastic processes. It is, therefore, a function of temperature and time. This is shown in Figure 7.28. At high temperatures, polymer chains are flexible and more rubber-like, and Poisson's ratio is rather high. Below glass transitions, less free volume exists and μ' gets smaller.

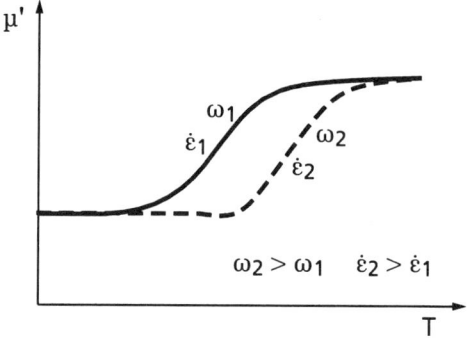

Fig. 7.28. Schematic plot of the Poisson's ratio versus temperature at different frequencies ω or strain rates $\dot{\varepsilon}$.

The exact measurement of μ' is difficult. In tensile specimens the cross sectional strain distribution is not very homogeneous, not even in the gauge length. One method is the measurement of the transverse and longitudinal sound velocities, v_t and v_l, respectively. The following relation can be used for the determination of μ'

$$\mu' = 0.5 \frac{(v_l/v_t)^2 - 2}{(v_l/v_t)^2 - 1} \tag{7.28}$$

Several results have been compiled in Figure 7.29. It can be seen that μ' increases at higher temperatures since more free volume gets available. At very low temperatures, μ' ranges from 0.29 to 0.39. It is instructive to compare the Poisson's ratio of the sequence HDPE, PTFE and PVFE, where the atoms of the side substituents become successively larger. (PVFE: polytrifluorochloroethylene).

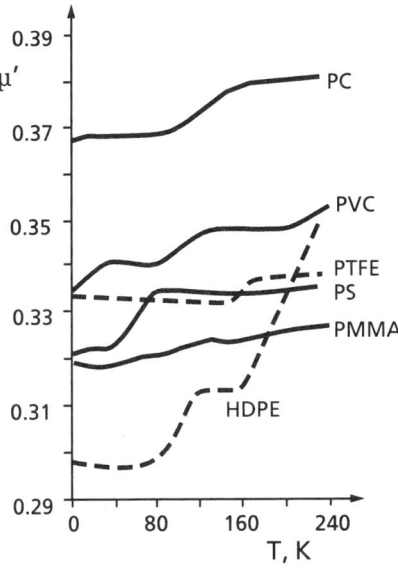

Fig. 7.29. Poisson's ratio versus temperature [7.10].

The results are:

$\mu' = 0.30$ for HDPE
$\mu' = 0.33$ for PTFE
$\mu' = 0.37$ for PVFE

Some explanation can be found in a tighter packing of HDPE which has smaller side substituents than PVFE. The bulky chloride atoms of PVFE may act as spacers and are thus responsible for more free volume [7.10].

7.11 ORIENTED POLYMERS

Self-orienting liquid crystalline polymers (LCPs), stretched polymers or polymer fibers are characterized by a large anisotropy. The analytical treatment of moduli is similar to that of fiber composites [7.11]. The modulus in the direction of orientation depends on the degree of orientation and on the type of polymers.

Polymer Fibers. Important polymer fibers are Kevlar fibers, which consist of aligned aramide molecules. Their longitudinal elastic modulus $E_\parallel$ is very large and nearly temperature-independent. The transverse modulus is larger than expected for aligned polymers. This is because not only Van der Waals forces control the interchain bond. The chemical structure is made such that strong hydrogen bridges increase transverse mechanical properties. The transverse modulus $E_\perp$ depends on temperature.

$E_\parallel \approx 150$ GPa; $E_\perp \approx 15$ GPa (77K) and $E_\perp \approx 10$ GPa (RT)

Liquid Crystalline Polymers (LCPs). Properties are discussed of the fully oriented LCP (Vectra A 950; similar results have been found for Vectran Hs 1000). Some anisotropy exists for the tensile moduli. The longitudinal modulus is rather small at RT because of slipping: $E_\| \approx 10 \text{GPa}$. On account of brittleness the transverse modulus is difficult to measure. Ultrasonic measurements give $E_\perp(293\text{K}) \approx 0.2$ GPa; $E_\perp(77\text{K}) \approx 0.4$ GPa.

The dependence on orientation is different for shear loading or torsion. The shear modulus G' arises from interchain shear forces, which arise from Van der Waals potentials (dashed lines in Fig.7.30). The solid lines represent molecular chains. It is interesting to note that the shear moduli at 0° and 90° orientations are nearly equal. The influences of shear forces by the Van der Waals potentials are similar. At 45° orientation the shear modulus is highest (about twice that at 0° or 90° orientations).

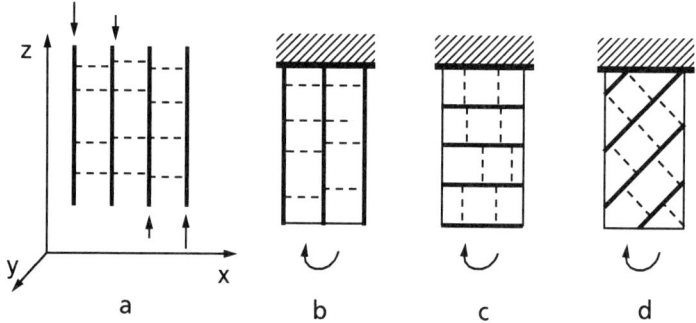

Fig.7.30: a) Shearing of oriented polymer chains.
b) Torsion about an axis parallel to the orientation.
c) Torsion about an axis perpendicular to the orientation.
d) Torsion at 45°.

Finally, a remark should be added on the Poisson's ratio μ of oriented polymers. For isotropic materials μ is between 0.25 and 0.5. For oriented anisotropic materials this is no longer true. Large strain in x- (or y-) direction, e.g., causes only a very small contraction of the rigid molecules in z-direction: $\mu_{xz} \approx 0.1$.

7.12 SUMMARY

(1) The deformation behavior is described by moduli or compliances under strain- and stress-controlled loading, respectively. If viscoelastic processes are involved, these loading types yield slightly different results. Below 150K or at very high frequencies, this distinction is negligible.

(2) The moduli of polymers are composed of fast components (originating from binding forces) and slow components (from viscoelastic place changes).

(3) At very low temperatures or with very short loading times (high frequency or high load rates), the unrelaxed modulus can be measured. The unrelaxed modulus is the linear-elastic modulus (fast component).

(4) In the vicinity of a secondary or tertiary glass transition temperature (dispersion region) the modulus (or compliance) is temperature- and time-dependent. The dependence on frequency or load rate, however, is much weaker than on temperature.

(5) The deformation behavior includes the following elements:
- elastic response : reversible, nondissipative, linear;
- viscoelastic response : reversible, dissipative, linear or nonlinear;
- viscous response : irreversible, dissipative, nonlinear.

At very low temperatures, elastic or linear viscoelastic behavior prevails. Nonlinear viscoelastic response is pronounced for thermoplastic polymers at elevated temperatures. Viscous flow is small or negligible below 100K and not too high loads. This is true especially for cross-linked polymers.

(6) Several types of moduli (or compliances) are defined for different time profiles of loading:
- load step: relaxation modulus
- cyclic loading: complex modulus
- constant load rate: secant and initial modulus

The moduli can be correlated by the superposition principle.
The moduli, except the secant moduli, are defined at low load amplitudes.

(7) For cyclic loading with angular frequency ω a maximum of the damping component and a modulus step occur at $\omega \tau = 1$.
The relaxation time τ is slightly different for stress- or strain-controlled loading above 100K and low frequencies.

(8) At very low temperatures the moduli E' (uniaxial) and G' (shear) of polymers, respectively, approach narrow bands of values, which are rather independent of the chemical composition. However, the crystalline fraction or the cross-link density exert some influence; highly crystalline or cross-linked polymers exhibit higher ranges of moduli.

(9) The Poisson's ratio depends on temperature and time, especially in the dispersion region. At 4.2K, the values of isotropic polymers range between 0.29 and 0.39. At glass transitions, the values increase since the free volume becomes larger.

(10) At high load amplitudes the stress - strain diagrams become nonlinear, even at 77K. Some viscous components remain after unloading.

(11) The stress - strain diagrams at high tensile load amplitudes become more linear and hysteresis effects become smaller after repeated load cycles. This behavior is important for fatigue cycling.

(12) Oriented polymers show an anisotropy of the deformation behavior. Well-known examples are Kevlar and LCPs.

7.13 REFERENCES

7.1 Schwarzl, F.R.; Mechanik der Polymere; p. 136; Springer Press; Berlin-Heidelberg (1990)

7.2 Pauling, L. and E.B. Wilson; Introduction to Quantum Physics; McGraw Hill-Verlag (1935).

7.3 McCrum, N.G., B.E. Read and G. Williams; Anelastic and Dielectric Effects in Polymeric Solids, John Wiley and Sons; p. 130.

7.4 Ref. 7.3; p.137

7.5 Ref. 7.3; p.234

7.6 Döll, W.; in: Advances in Polymer Sci. 52/53,p.106; Springer Press; Berlin-Heidelberg (1983).
7.7 Hartwig, G.; B. Kneifel and K. Pöhlmann; in Advances in Cryogenic Engineering (Materials), Vol. 32; p. 169; Plenum Press, New York, (1986)
7.8 Hartwig, G.; Habilitationthesis, University Erlangen (1989).
7.9 Haward, R.N. in: The Physics of Glassy Polymers; p. 330 and p. 375; Applied Science Publishers Ltd, London (1973)
7.10 Perepechko, J.; Low-Temperatures Properties of Polymers; p. 241 Pergamon Press, MIR Publishers, Moscow (1980).
7.11 Hartwig, G.;Cryogenics, Vol.28; p. 220.

General Reading

1. Ferry, J.D.; Viscoelastic Properties of Polymers; (3. Edition); John Wiley and Sons, New York (1980).
2. Advances in Polymer Science 52/53); Crazing in Polymers; Editor: H. H. Kausch; Springer Press; Berlin-Heidelberg (1983).
3. Advances in Polymer Science 91/92, Vol. 2; Crazing in Polymers; Springer Press; Berlin-Heidelberg (1990).

DIELECTRIC PROPERTIES AND THEIR CORRELATIONS

Contents

8.	Dielectric properties and their correlations	175
8.1	General introduction	175
8.2	Definitions	176
8.3	Permittivity and compliance	177
8.4	Dielectric and mechanical damping at low frequencies	178
8.5	Results at high frequencies	182
8.6	Permittivity and its asymptotic behavior	184
8.7	Calculation of dipole moments	184
8.8	Summary	185
8.9	References	186
Appendix		
8.A	Data	

8. DIELECTRIC PROPERTIES AND THEIR CORRELATIONS

8.1 GENERAL INTRODUCTION

There are several similarities of material reactions, when activated mechanically or electrically. It concerns moduli and permittivity or damping behavior. The following parameters or properties are analogous:

Mechanical case	Dielectric case
Stress (Pa)	Voltage
Strain	Dielectric displacement
Damping	Resistance (Ω)
Spring	Capacitance
Complex compliance F^* or J^*	Complex permittivity ε_e^*
Storage compliance F' or J'	Permittivity ε_e' (real part)
Storage moduli E' or G'	$1/\varepsilon_e'$
Damping moduli E'' or G''	$1/\varepsilon_e''$
$\tan\delta_m = E''/E' = G''/G'$	$\tan\delta_e = \varepsilon_e''/\varepsilon_e'$

There are, however, differences between the dielectric and mechanical properties as regards the dependence on the dipole moments and the relative contributions by electron components. The mechanical loads are transmitted by binding forces, which depend only weakly on the dipole moments of polymers. Electric fields act directly on molecular groups which possess permanent or induced dipole moments. The mechanical deformations are therefore rather independent of dipole moments. The dielectric properties, by contrast, are strong functions of the dipole moments of the polymer. Both mechanical loads and electric fields deform the electron configurations of a material and cause place changes of mobile groups in polymers. The relative contributions of both processes are different, however, for the mechanical and electric case. Molecular place changes by jumping or tunneling processes in amorphous polymers can be activated by mechanical loads and electric fields. Therefore, a correlation of mechanically and electrically activated properties is expected to exist. The models of viscoelastic or tunneling place changes can be used for mechanical as well as electrical activations. An advantage is that dielectric measurements allow very high frequencies to be applied. At low temperatures resonant tunneling is, for example, a domain of high frequencies. This will be considered in Subsection

8.5.2. In the Sections 8.3 and 8.4 the correlation of compliance and permittivity and the damping behavior will be treated. For introduction a schematic representation of the permittivity and dielectric loss-factor $\tan\delta_e$ of a polymer is given in Figure 8.1; [8.1; 8.2].

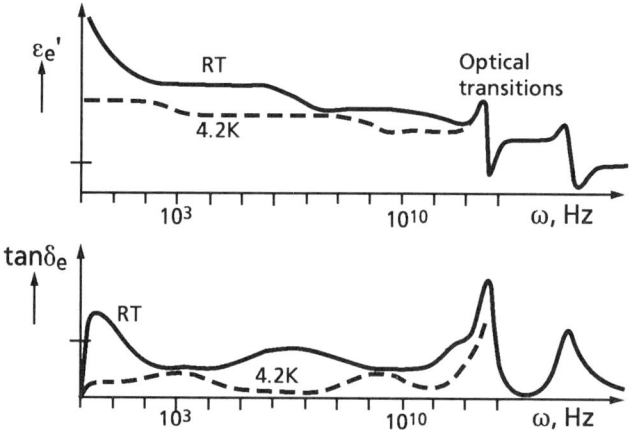

Fig. 8.1. Schematic representation of the permittivity ε_e' and the loss-factor $\tan\delta_e$ versus frequency at 4.2K and 300K. (Optical transitions occur above 100GHz).

The frequency- or temperature dependence of ε_e' and $\tan\delta_e$ is small except in the low-frequency or optical region. It depends on the polarity of polymers. Below 20K, dielectric damping of amorphous polymers is dominated again by tunneling processes. A plateau region at very low temperatures has been found [8.3].

8.2 DEFINITIONS

The relative permittivity ε_e' of an insulator is defined by the ratio of the capacitance C with a polymer and a vacuum between the electrodes. An important dielectric quantity is the complex permittivity, defined by the permittivity ε_e' and the damping permittivity ε_e''

$$\varepsilon_e^* = \varepsilon_e' - i\varepsilon_e'' \qquad (8.1)$$

This definition is analogous to that of the complex mechanical compliance. The imaginary damping part $i\varepsilon_e''$ has a negative sign (see negative phase angle of Fig. 7.14). The loss-factor is defined by

$$\tan\delta = \varepsilon_e''/\varepsilon_e' \qquad (8.2)$$

The permittivity and the loss-factor are controlled by the dielectric polarization, which is defined as the density of dipole moments per volume. For polymers the mean dipole moment $\bar{\mu}$ of a repetitive unit is used (see Section 8.7). The unit of a dipole moment is a Debye:

$$1D = 3.34 \cdot 10^{-30} \text{ Coulombmeter.}$$

8.3 PERMITTIVITY AND COMPLIANCE

Mechanical load effects two types of reaction in polymers: deformation of the electron configuration mainly due to Van der Waals potentials (fast component, elasticity) and activation of viscoelastic or tunneling place changes of molecular elements (slow component). In an analogous way, an electric field polarizes electron configurations (fast component) and induces slow orientation effects of polar chain segments or side groups. Thus, both types of activation act on fast and on slow components:

Mechanical activation	Dielectric activation
Fast electron deformation of electron configurations	Fast electron polarization
Slow viscoelastic deformation (weakly depending on the dipole moment)	Slow orientation polarization (strongly depending on the dipole moment)

The electron components are nearly independent of temperature and frequency up to ≤10GHz. Place changes or orientations of bulky molecular segments, by contrast, are related to long relaxation times and are strong functions of frequency and temperature.

As already mentioned there is a great difference between mechanical and dielectric activations: The mechanical deformation behavior is not much dependent on the polarization of a polymer, whereas the orientation polarization by electric fields depends on the permanent or induced dipole moments of relaxing molecular groups. Thus, the orientation polarization is different for polymers having high or low dipole moments. For viscoelastic mechanical deformations the dependence is not so marked. Only part of the Van der Waals binding forces which transfer mechanical load originate from dipole-dipole interactions; the main part is due to nonpolar dispersion forces (see Section 1.2). This difference in mechanical and dielectric behavior is plotted schematically in Fig.8.2. The permittivity ε_e' and the shear compliance J' are considered to be analogous properties. The step height of ε_e' at tertiary or secondary glass transitions T_g is greater for polar groups than for low-polar ones. The steps of moduli or compliances are generally different from that of the permittivity, but they occur at the same temperature-frequency position.

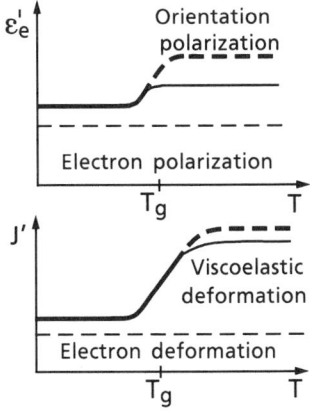

Fig. 8.2. Schematic plot of the permittivity ε_e' and the mechanical compliance J' versus temperature.
—— low-polar
– – – high-polar

Experimental values of ε_e' and J' are given in Figure 8.4. The ratio ε_e'/J' is plotted in Figure 8.3. It is different for the investigated polymers since the dipole moments of their relaxing segments are different [8.4].

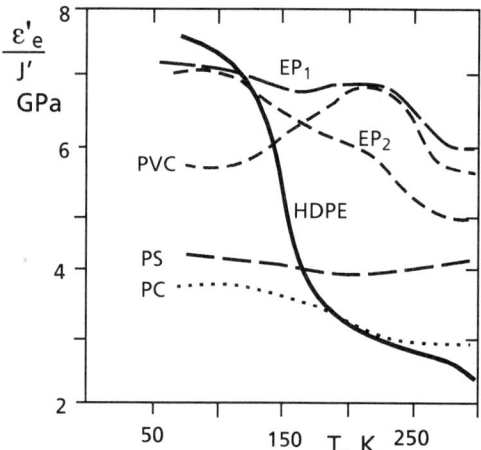

Fig. 8.3. Ratio of ε_e'/J' versus temperature. Frequency 7-10Hz.
EP_1: rigid epoxy resin EP_2: flexibilized

For PS both the compliance and the permittivity are rather independent of temperature (no glass transitions) and the same applies to their ratio. EP and PVC are polar polymers; the step-like changes of J' and ε_e' are similar and the ratio stays rather constant. HDPE, by contrast, is rather nonpolar. The strong secondary glass transition ($\approx 160K$) has therefore a much smaller effect on ε_e' than on J'. Since the compliance increases strongly with temperature the ε_e'/J' ratio decreases (see Fig. 8.4a). The dipole moments $\bar{\mu}$ of the polymers considered in Figure 8.3 differ by a factor of 5 while ε_e'/J' changes by less than a factor of 3. This means that the ε_e'/J' ratio is not strongly sensitive to $\bar{\mu}$. The ratio of dielectric and mechanical damping, by contrast, is strongly dependent on $\bar{\mu}$.

8.4 DIELECTRIC AND MECHANICAL DAMPING AT LOW FREQUENCIES

The dielectric damping behavior is not influenced by the electronic components but is determined by the orientational polarization. Polymers with a high dipole moment yield a higher loss-factor $\tan\delta_e$ than nonpolar ones. Experimental results of the dielectric loss-factor $\tan\delta_e$ and the permittivity ε_e' of several polymers are plotted in Figures 8.4a to 8.4f. As already mentioned, a correlation of dielectric properties and their mechanical counterparts is expected to exist. Therefore, the mechanical loss-factor $\tan\delta_m$ and the shear compliance J' are also shown (J' is considered to be representative of the mechanical case; the behavior is similar for the compliance F' under tensile loading).

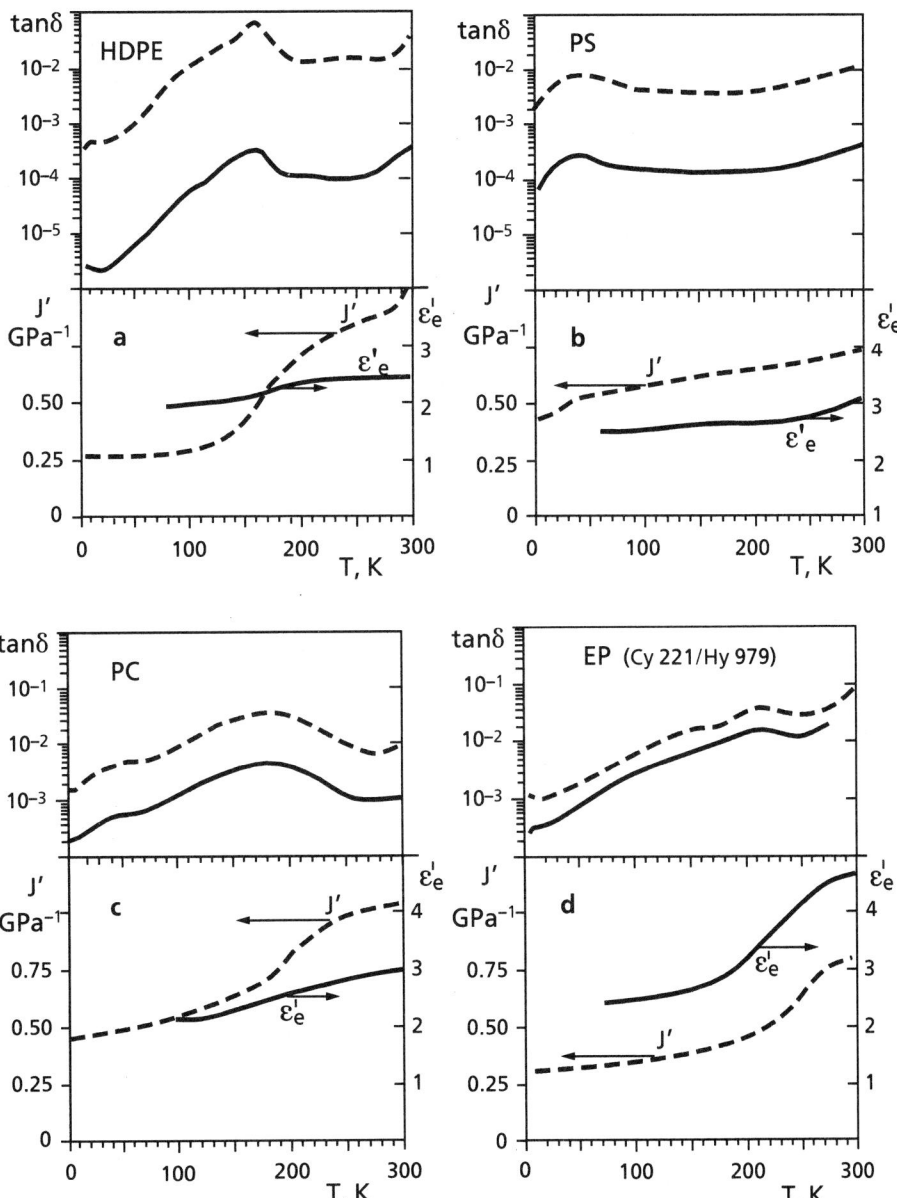

Fig. 8.4. a) to d) Loss-factors $\tan \delta_e$ and $\tan \delta_m$, permittivity ε_e', and shear compliance J' versus temperature. – – – mechanical properties; ——— dielectric properties

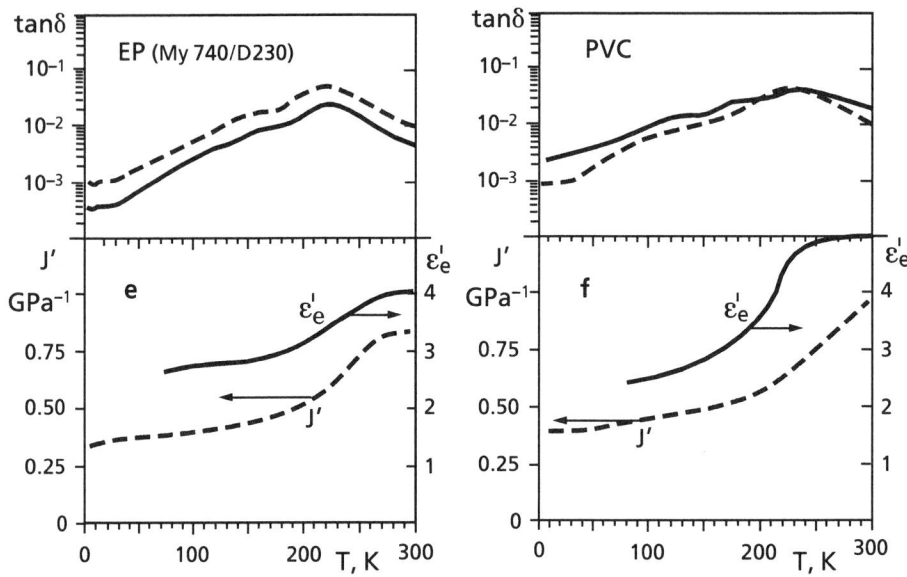

Fig. 8.4. e) and f): Loss-factors tan δ_e and tan δ_m, permittivity ε_e', and shear compliance J' versus temperature. − − − mechanical properties; ——— dielectric properties

Several similarities in the dielectric and mechanical behavior become evident: The first result is that the frequency-temperature positions of the damping maxima are equal for the mechanical and dielectric activations. This means that the relaxation times are the same for both types of activation:

$$\tau_e = \tau_m \tag{8.3}$$

The second result is evident from the semilogarithmic plot in Figures 8.4a to 8.4f. The ratios of both loss-factors are independent of temperature (at least below 250K) and differ by a constant factor f for each polymer. Numerical values of the ratio f are shown in Fig.8.5.

$$\frac{\tan\delta_e}{\tan\delta_m} = f \tag{8.4}$$

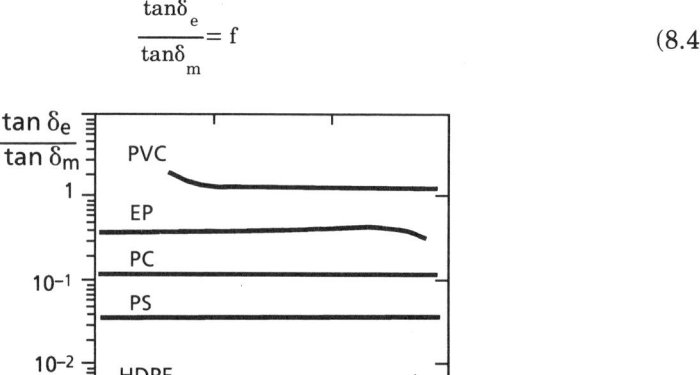

Fig. 8.5. Ratio of dielectric and mechanical loss-factors versus temperature. Frequency 7-10Hz.

DIELECTRIC PROPERTIES AND THEIR CORRELATIONS

The third result concerns the value of f. The polymers investigated differ by their mean dipole moments $\bar{\mu}$. HDPE is rather nonpolar and the dielectric loss-factor is much smaller than the mechanical one ($f < 10^{-2}$). The factor f is smallest for HDPE and gets larger for polymers with increasing mean dipole moments $\bar{\mu}$. For PVC which is a rather polar polymer the values of dielectric- and mechanical loss-factors are similar ($f \approx 1$). In general, the ratio f of loss-factors seems to be a function of $\bar{\mu}$. In Figure 8.6 the ratio f is plotted versus $|\bar{\mu}|$. A double logarithmic plot yields a straight line, which obeys roughly the empirical relation [8.4]

$$f \equiv f(\bar{\mu}) = \frac{\tan \delta_e}{\tan \delta_m} \approx |\bar{\mu}|^{2.4} \quad (\bar{\mu} \text{ in Debye units})^{1)} \tag{8.5}$$

($|\bar{\mu}|$ means its absolute value for cases when μ is negative by definition; see Section 8.7). The value of $|\bar{\mu}|$ covers only the components of the mobile groups of a polymer. For example, only the wagging of the phenyle rings of PS causes damping below RT. Their bond moment determines $|\bar{\mu}|$ in Eq.8.5. For most other polymers, however, segmental motions similar to crankshaft motions prevail, and it is a good approximation to use for $|\bar{\mu}|$ the mean dipole moment of a repetitive unit. The dipole moment of a repetitive unit can be calculated for polymers of a simple composition. Examples will be given in Section 8.7.

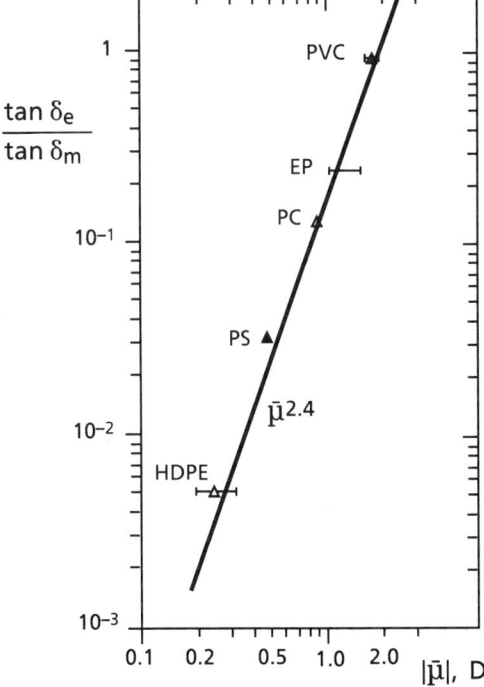

Fig. 8.6. Ratio $\tan\delta_e/\tan\delta_m$ versus the mean dipole moments $|\bar{\mu}|$ of various polymers.

The other way round, this relation can be used to determine $\bar{\mu}$ of polymers with a complicated chemical composition. The ratio f can easily be measured at any temperature, and Eq.8.5 gives $\bar{\mu}$, of the mobile group, which in most cases is represented by the repetitive unit.

---

[1] Kirkwood found a similar relation $\varepsilon_e'' \propto \mu^2$ [8.5].

8.5 RESULTS AT HIGH FREQUENCIES

Investigations in the very high frequency region are feasible due to dielectric excitations. Two main topics will be considered:

- Loss-factor and permittivity
- Tunneling processes at high frequencies.

An example of the frequency dependence of the dielectric loss-factor $\tan\delta_e$ is shown in Figure 8.7 for MDPE at 4.2K and 300K. It is obvious that the frequency dependence is rather weak. (MDPE is the polymer with the lowest known dielectric loss-factor). Polymers with higher polarities show a somewhat stronger frequency dependence [8.1; 8.2].

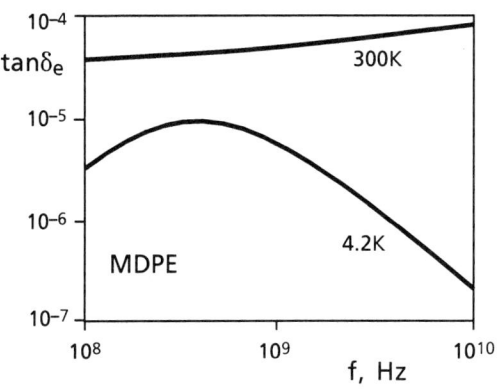

Fig. 8.7. Dielectric loss-factor of middle density polyethylene (MDPE) versus frequency.

8.5.1 Relaxation Tunneling

Damping by relaxation tunneling occurs at dielectric high-frequency excitations as well. The low-temperature behavior of $\tan\delta_e$ at three constant frequencies is plotted for HDPE in Figure 8.8. Also a plateau region exists for dielectric excitations. At high frequencies it is shifted to higher temperatures. The plateau height, however, is not very sensitive to frequency. It varies within a small range when the frequency f is varied by 8 orders of magnitude. The plateau level is not a strong function of frequency, but it stongly depends on the dipole moments of the tunneling elements.

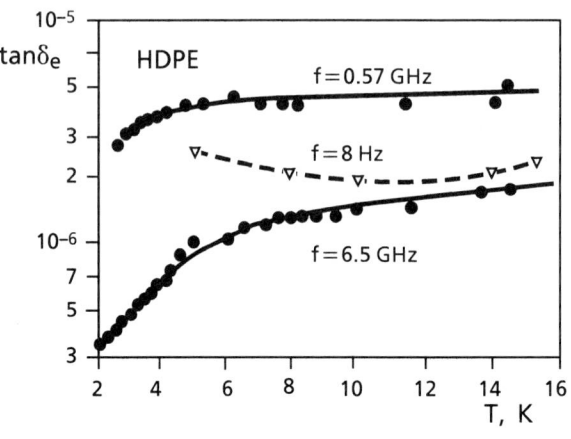

Fig. 8.8. Loss-factor $\tan\delta_e$ versus temperature at three frequencies [8.3].

DIELECTRIC PROPERTIES AND THEIR CORRELATIONS

Below the plateau region the drop of $\tan\delta_e$ at 6.5 GHz already starts at 5K, whereas it starts at about 0.1K at low frequencies (see the mechanical analog in Fig. 6.18). The drop of $\tan\delta_e$ below 4.2K is plotted in Figure 8.9 for several polymers having different dipole moments and a different crystalline content. Highly crystalline polymers show small loss-factors since only the amorphous phase is dissipative. PVC, which is amorphous and polar, exhibits the highest values in this figure. MDPE shows the lowest values; it is semicrystalline and nonpolar. The large difference of $\tan\delta_e$ for low-, high-, and middle density polyethylene is not quite clear.

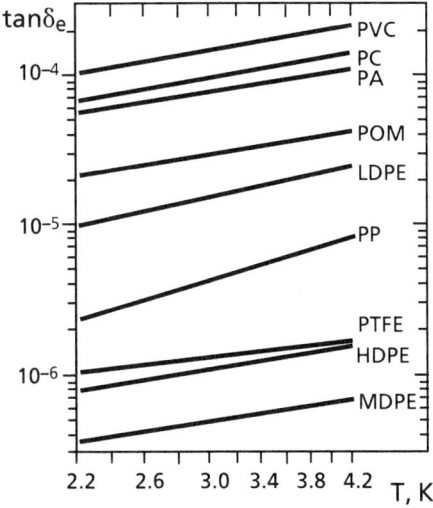

Fig. 8.9. Loss-factor of several polymers at 6.5GHz below the plateau region [8.1; 8.2].

The dielectric loss-factor strongly depends on polarity even at low temperatures, and a very broad band of values has been found for polymers. This is different from mechanical damping behavior.

8.5.2 Resonant Tunneling

Resonant tunneling occurs when a phonon is absorbed, which fits to the difference E in energy levels of a tunneling system ($E = \hbar\omega$). This process requires phonons with much higher energies (frequencies ω) than those for relaxation tunneling, which only modulate the energy levels (see Section 6.5.2). At low frequencies resonant tunneling occurs at extremely low temperatures. It can be shifted to reasonable temperatures by applying very high frequencies (e.g.,10GHz). This can be done by electric waves, which couple to the dipole moments of molecular groups and cause vibrations (phonons).

By tunneling absorption processes the population of the ground level is shifted to the upper level of the two-level system till saturation occurs. Therefore, resonant tunneling absorption depends on the activation intensity I. Results are given in Figure 8.10 for glass (Suprasil). It can be seen that the damping values decrease with an increased intensity of activation.

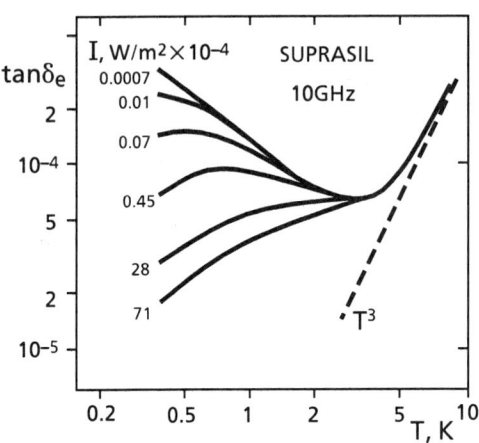

Fig. 8.10. Loss-factor due to resonant tunneling at high frequencies (10GHz) and different intensities I. The dashed line marks the T^3 drop of relaxation

8.6 PERMITTIVITY AND ITS ASYMPTOTIC BEHAVIOR

The permittivity of polymers covers a range of $\varepsilon_e' \approx 2$ to 5 (see Fig. 8.4). At low temperatures or at high frequencies nearly all contributions stem from the electronic components, which are not influenced by polarity. The values of ε_e' at very low temperatures approach a narrow band between $\varepsilon_e' \approx 2$–2.7, independent of the chemical compositions of polymers. In Appendix 8 A numerical values of ε_e' have been compiled for several polymers at different frequencies.

8.7 CALCULATION OF DIPOLE MOMENTS

Dipole moments of rigid molecules can be determined experimentally in solutions or in the gaseous state. Several monomers constituting polymers have been determined and can be used to calculate the mean dipole moments $\bar{\mu}$ of polymers by adding vectorially [8.6].

$$\mu = \Sigma \, \vec{\mu}_j \quad \text{(Debye units } D\text{)} \tag{8.6}$$

The basic components are the bond moments; some of them will be listed in Table 8.1 [8.7].

Table 8.1: Bond moments

C – F	1.39 D
C – Cl	1.47 D
C = O	2.4 D
C = N	1.4 D
C – phenyl ring	0.37 D
C – H	– 0.29 ± 0.1D (negative direction)

The direction or sign of the bond moments is taken from the negative to the positive charge center. The (negative) bond moment of C – H is influenced by the neighboring side atoms, e.g., by F− or Cl atoms (this is included in the error limits of $\bar{\mu}$ (C−H)). The tetrahedal configurations of the bond moments lead to partial cancellations of bond moments. A valence angle of 109° is assumed. Some simple examples will be given:

1) For PE the dipole moment of a monomer, CH_2, is given by the projections of the bond moments C−H. The (negative) bond moments $\bar{\mu}(C-H)$ span an angle of 109°:

$\bar{\mu}$ (C−H_2) = −2 $\bar{\mu}$ (C−H) cos 54.5 = −2·0.29·0.45 ≈ −0.27 D.

2) For PVC the main contribution is made by the monomer CHCl. It can be calculated from the known value of $\bar{\mu}$ (CH_3Cl) = 1.87D, and vectorially reduced by the (negative) values of two C−H bond moments [8.6].

$\bar{\mu}$(CHCl) = $\bar{\mu}$ (CH_3Cl) − 2 $\bar{\mu}$ (C-H) cos (180 − 109) = (1.87−2·0.3·0.33) =
= 1.67 ± 0.1D

3) For PS the main contribution is given by the phenyl group.

For these estimations it has been assumed that the neighboring CH_2 monomers are atactically distributed and do not make effective contributions. In a rough approximation the values of the monomers can be used for a repetitive unit; e.g., [CH_2–CH_2]; [CH_2–CHCL]; [CH_2–CH◯]. The following dipole moments are valid for the polymers considered:

$\bar{\mu}$ (PE) :   − 0.27 ± 0.1 D
$\bar{\mu}$ (PS) :   0.40 ± 0.04 D
$\bar{\mu}$ (PC) :   0.85   D [8.7]
$\bar{\mu}$ (EP) :   1.2 ± 0.2 D
$\bar{\mu}$ (PVC) :   1.67 ± 0.1 D

8.8 SUMMARY

There are several correlations and similarities between dielectric and mechanical properties of polymers.
(1) Dielectric and mechanical excitations cause deformations of the electron configurations (fast component) and place changes (or orientations) of molecular segments or side groups (slow component).
(2) The permittivity is constituted of frequency and temperature-independent electron components and contributions from orientational polarization, which depend on frequency and temperature and on the dipole moment of the mobile groups. The permittivity therefore generally shows different step heights at (secondary or tertiary) glass transitions than the analog mechanical compliance, which is not very sensitive to polarization. The permittivity of nonpolar polymers is little dependent on temperature or frequency.
(3) The relaxation times of the slow components are equal for dielectric and mechanical activations.
(4) The dielectric loss-factor $\tan\delta_e$ depends strongly on the dipole moments of the mobile molecular groups. This is different from the mechanical case.

Therefore, dielectric loss-factors cover a much larger range than mechanical ones.

(5) The ratio of dielectric to mechanical loss-factors is independent of temperature, but depends heavily on the dipole moment $\bar{\mu}$ of the polymer. It holds: $\tan\delta_e/\tan\delta_m \approx |\bar{\mu}|^{2.4}$ ($|\bar{\mu}|$ means the absolute value).

(6) The ratio of permittivity ε'_e and mechanical compliance depends on the dipole moment and the temperature or frequency.

(7) A pronounced temperature and frequency dependence of $\tan\delta_e$ exists only at low frequencies ($\ll$MHz). In the GHz region and at low temperatures dielectric damping becomes very small.

(8) Tunneling effects in dielectric excitations occur as well. A plateau region from relaxation tunneling has been found. The plateau height, however, depends on the polarity.

(9) Resonant tunneling processes occur at extremely low temperatures. It can be shifted to reasonably high temperatures by applying very high frequencies (e.g., 6GHz at 5K). This is a domain of dielectric excitations. High - frequency phonons are created by electric waves. Resonant tunneling occurs by absorption of phonons, whose energy fit the energy difference of a two-level tunneling system. The absorption depends on the phonon (photon) intensity I because of saturation effects.

References

8.1 Meyer, W.; Proc. 6. Intern. Cryogen. Eng. Conf.; May 1976, Grenoble.

8.2 Meyer, W.; Solid State Communications, Vol. 22 (1977), p. 285-288 Pergamon Press.

8.3 Schickfus, v. M. and Hunklinger, S.; Physics Letters, Vol. 64 A, (1977) p. 144.

8.4 Hartwig, G. and Schwarz, G. in Advances in Cryogenic Engineering, Vol. 30 (1984), pp. 61-70; Eds. A. Clark and R.P. Reed; Plenum Press, N.Y.

8.5 Kirkwood, J. in Hedwig, P. : Dielectric Spectroscopy of Polymers, (1977), p. 309.

8.6 Hedwig, P.; Dielectric Spectroscopy of Polymers, (1977), p. 18.

8.7 Krevelen, van D.W.; Properties of Polymers, (1972), Elsevier Publishing Company; p. 212.

FRACTURE BEHAVIOR OF POLYMERS

Contents

9.	Fracture behavior of polymers	189
9.1	Introduction and survey	189
9.2	Fracture mechanics	192
9.3	Measurements of K_{IC} and G_{IC}	198
9.4	Processes prior to fracture	201
9.5	Dependence on deformation rate and temperature	205
9.6	Fatigue strength	211
9.7	Summary	217
9.8	References	217
Appendix		
9A	Data	

9. FRACTURE BEHAVIOR OF POLYMERS

9.1 INTRODUCTION AND SURVEY

Fracture analysis is complicated for polymers since, besides of temperature and time dependence, there are involved effects from plastification, chain orientation and adiabatic temperature rise. It is the aim of this section to describe several fracture processes which are specific of polymers. Fracture behavior is determined **experimentally** by the mode and time profile of loading. Stress- and strain controlled loading yields different behavior as well as loading in tension, compression or shear (torsion). Compressive strength is higher than tensile strength; shear strength is the lowest one.

Different time profiles of loading cause different fracture behavior. Of technical interest are the following cases:

- **Load ramp to critical load.**
  Fracture strength and strain as well as the fracture process involved depend on the time profile of loading, especially on the strain- or stress rates if viscous or viscoelastic processes are involved.

- **Long-term constant subcritical load.**
  The time to failure depends on the height of the load step. Creep- or relaxation processes determine the fracture behavior at loads well above the linear viscoelastic range (This fracture mode is called static fatigue).

- **Cyclic subcritical load.**
  The fatigue life depends on the load amplitudes and the frequency. The upper and the lower load levels determine the fatigue life.

Environment, e.g., a LN_2 bath, may have some influence on fracture processes (stress corrosion, enhanced craze formation).

Fracture strength as a **material property** is determined by the bond strength, the induced fracture processes and the stress distribution within a body. **The bond strength** within and between polymer chains depends on temperature. Due to thermal contraction a tighter packing and thus a higher bond strength at decreasing temperatures results. At very low temperatures chains or chain atoms have nearly no thermal energy capable of counteracting the binding potentials. **Induced fracture processes** are crazing, necking or plastification. **Inhomogeneous stress distributions** lead to stress concentrations which reduce fracture strength.

9.1.1 Stress Concentrations

From a micromolecular point of view it is astonishing that isotropic polymers exhibit only a poor fracture strength, despite the fact that polymer chains are extremely strong. Binding structure of most polymer backbones would allow an enormous fracture strength (up to 15GPa). Entangled, isotropic polymers exhibit a tensile strength in the range of 30 to 200MPa only. The upper value applies to low temperatures. The reason for a low fracture strength arises because of inhomogenieties of the material structure which cause stress concentrations.

Molecular Stress Concentration. Entanglement of polymer chains is a source of stress inhomogeneities. Polymers carry high load in the chain direction only. Perpendicular only small strength is provided by the Van der Waals potential. Several load elements are sketched in Figure 9.1 for a glassy polymer. The chains are assumed not to undergo conformational changes at low temperatures.

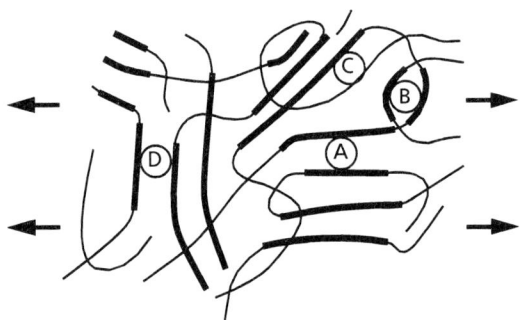

Fig. 9.1. Load components within entangled polymer chains
(thick lines are drawn only for demonstration of components).

Most of the load carrying capacity is given by components A of chain segments in load direction, but they exist in a low density only. Entangled elements B are also strong, but they will fail prior to A elements. The off-load axis elements C carry gradually more load when A- and B elements break. The weakest elements D are Van der Waals-bonded chains perpendicular to the load direction. The unbalanced and inhomogeneous stress concentration within entangled chains reduces tensile strength considerably. Compressive strength, by contrast, is much less influenced by the molecular entanglement. Oriented polymers, e.g., fibers, exhibit a different behavior. A high tensile strength is concentrated in fiber orientation (up to 2GPa), the compressive strength, however, is small (buckling effect).

Microstress Concentration. Microcracks, voids, crazes or notches exist in each polymer. When loaded, stress concentration at their crack tips occur, and they become nuclei of macroscopic cracks. Analytical treatment is provided by fracture mechanics. Assuming a homogeneous structure (neglecting the chain structure) the stress distribution at a crack tip can be calculated. The decisive parameter is the stress intensity factor K, which depends on the mode of loading (tension, shear).

Thermal Stresses. When polymers are cooled too quickly, inhomogeneous thermal stress gets frozen, and reduces tensile strength. Thermal cycling accumulates this effect within certain limits.

9.1.2 Crack Initiation and Propagation

Local overloadings, are described by stress intensity factors. They are the static aspect of fracture initiation. The stress intensity factor can be calculated from the overall load, the given crack- or craze length and the sample geometry. This property usually is measured with fracture mechanical specimens. The dynamic aspect investigates the crack propagation, which is controlled by the energy balance of the released energy rate G and the consumed energy rate R, the fracture energy. Both are related to the newly created crack area of a fracture process. (The energy concept more generally allows to analyse real and virtual crack propagations). The fracture energy rate R, in addition, is influenced by local heating ΔT and plastification in the crack vicinity which depends on the crack velocity à. An example of the interplay of G and R during crack propagation is given in Section 9.2.4.

9.1.3 Deformations and Fracture

Fracture processes of polymers are strongly influenced by deformation- or yielding processes which depend on temperature and time. At very low temperatures no yielding is possible and fracture is brittle. In this case the fracture strength is relatively high, the strain to failure, however, is small. In the vicinity of RT or above, fracture strain is strongly increased by yielding or other inelastic deformation processes. For duroplastic polymers yielding is suppressed by cross-linkage. For several thermoplastic polymers (e.g., PC) yielding already occurs well below RT. On a macroscopic scale the interplay of deformation, yielding and fracture at different temperatures is reflected in the stress-strain behavior. In Figure 9.2 schematic stress-strain curves are plotted.

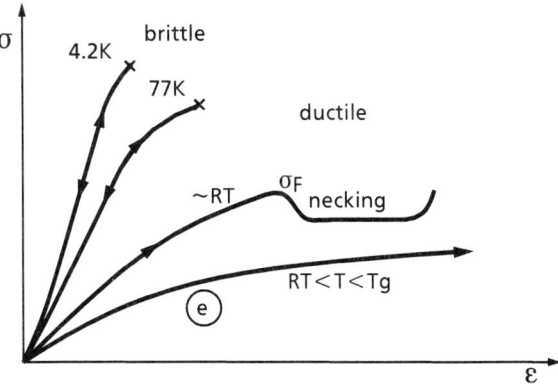

Fig. 9.2. Schematic stress-strain curves at different temperatures.

At 4.2K a nearly linear stress-strain behavior up to fracture is exhibited. Only small nonlinearities occur which, however, are reversible. The fracture strain is about 2% to 3.5%.
At 77K nonlinear contributions become larger but almost reversible processes prevail, and the fracture mode is brittle. The fracture strain is of the order of 3% to 8%. A deviation at high strain rates, which cause adiabatic heating, is treated in Section 9.5.

At RT and moderate strain rates almost yielding and necking (strain softening) of tensile specimens is observed. A peak of the curve in Figure 9.2 characterizes the yield stress σ_F. After stretching, an orientation between the necking zones occures which elongates the specimen length. This behavior is especially true for semicrystalline thermoplastic polymers. At elevated temperatures between RT and the main glass transition temperature T_g an overall yielding may take place without a peak in the strain curves (see curve (e) in Fig.9.2). Duroplastic polymers do not show the process of necking.

Fracture behavior also is influenced by microscopic yielding, which is enhanced at locations of stress concentrations, especially at crack tips. Crack tip zones are formed, whose properties can be influenced prior to the crack initiation and/or during the propagation. Heating of the crack tip zone occurs at high deformation rates or at fast running cracks. Orientation effects are increased at increased temperatures and slow deformation rates. An example of strong local orientation by yielding is the formation of crazes, whose crack tip zones consist mainly of aligned fibrilles.

9.2 FRACTURE MECHANICS

The fracture strength of a body depends on its geometry, its surface quality and inhomogeneities and on the type of loading. The crucial point is the nucleation and development of cracks. Crack formation and crack development at low temperatures can be described by "Linear Elastic Fracture Mechanics" LEFM which assumes linear elastic or linear viscoelastic behavior of the bulk material. Only locally in the vicinity of a −highly loaded− crack, nonlinearities and plastic deformations are allowed. Two concepts are applied:

- Crack initiation   → stress intensity factor.
- Crack propagation → energy balance between released and consumed energy.

Main emphasis of this chapter is put on fracture processes at low temperatures.

9.2.1 Stress Intensity Factor

Stress concentrations in the vicinity of cracks can be described by stress intensity factors K, which depend on the mode of loading; K_I for tensile loading. It is a function of the given crack length a, the overall stress σ and the correction factor Y(a/W) which takes into acount the sample geometry; W is the sample width. K_I is a parameter for describing the stress distributions [9.1]. It can be calculated from .

$$K_I = \sigma \, a^{1/2} \, Y(a/W) \quad (9.1a)$$

The overall stress σ is defined by the applied force P per total sample cross-section B·W

$$\sigma = \frac{P}{BW} \quad (9.1b)$$

where B is the sample thickness.

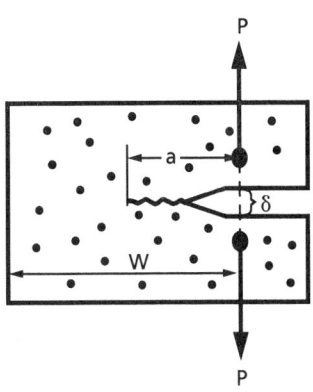

Fig. 9.3. Specimen for crack propagation measurement.
P: force
δ: crack opening displacement (COD)
a: crack length

The onset of crack development is determined by the critical value K_{IC}, which gives a correlation of the critical crack initiating stress σ_c and the crack length a. K_{IC} is a specific material parameter of a polymer. The critical stress σ_c to fracture and thus K_{IC} is a function of temperature, deformation rate (e.g., COD rate $\dot{\delta}$) and a preloading time Δt.

$$K_{IC}(T, \dot{\delta}, \Delta t) = \sigma_c(T, \dot{\delta}, \Delta t) \, a^{1/2} Y \tag{9.1c}$$

For a rectangular compact tension specimen (Fig.9.3) the correction function $Y = (2\pi)^{1/2}$ and therefore (see also Eq.9.4 b)

$$K_{IC} = \sigma_c (2\pi a)^{1/2} \tag{9.1d}$$

Several types of crack opening specimens are applied for measuring K_{IC}.

9.2.2 Energy Concept

The dynamic aspect of crack propagation is controlled by the balance of released and consumed energy.

Energy Release Rate G. During crack propagation, part of the elastically stored energy E_{el} is released. The energy release rate at tensile loading G_I (mode I) is defined by the released energy per newly created fracture area ΔA.

$$G_I = -\frac{\Delta E_{el}}{\Delta A} \tag{9.2a}$$

It is possible to calculate G_I at a real or virtual crack propagation, which allows statements of fracture criteria. Crack propagation starts at a critical value G_{IC}, which depends on the temperature, the COD rate $\dot{\delta}$ and a preloading time Δt.

$$G_{IC}(T, \dot{\delta}, \Delta t) = -\frac{\Delta E_{el}}{\Delta A} \tag{9.2b}$$

The value of G_{IC} can be determined from P-δ diagrams measured on crack opening specimens (see Figs. 9.3 and 9.7). G_{IC} is a material parameter [9.2].

Fracture Energy Rate R. Crack propagation consumes energy, the fracture energy rate R, which again is related to the newly created area ΔA. Several energy absorbing processes are involved

$$R = \frac{dE_s}{dA} + \frac{dE_{pl}}{dA} + \frac{dE_{vib}}{dA} + \cdots \tag{9.3}$$

For polymers most energy is consumed by plastification of the moving crack tip zones. The plastification energy E_{pl} is therefore high, the surface energy E_s is negligible, and the vibrational energy E_{vib} of broken chains is small.

The balance between released and consumed energy determines the behavior of an existing crack.

$G_I < R$ no crack propagation
$G_{IC} = R$ stable crack propagation
$G_I > R$ accelerated, unstable crack propagation

In the latter case the difference between G_I and R is balanced by a term of kinetic energy dE_{kin}/dA due to a crack acceleration [9.3]. For polymers at low temperatures mainly unstable crack propagation is realized [9.4]. The features of G_I and R are illustrated by an example given in the following section.

9.2.3. Crack Arrest Behavior

The interplay of G_I and R is responsible for the intrinsic crack arrest of unstable propagating cracks in polymers. Crack propagation experiments on crack opening specimens at low temperatures in most cases exhibits a unstable and discontinuous crack propagation. The values of G_I and R vary differently during crack propagation. Unstable cracks stop by themselves after a crack propagation length Δa. The reason arises from the interplay of G_I and R as a function of the crack length a. It can be shown that for Chevron specimens G_I decreases smoothly with the crack length a. The fracture energy rate R, by contrast, is a strong function of the crack length during fracture propagation. Figure 9.4 shows a model, which can help to explain the crack arrest behavior. The real experiment starts with an initial crack, which has been inserted by external load. This initial crack exhibits all the features, which determine the fracture process. If sufficient load is applied crack propagation starts when $G_{IC} = R$. While G_I decreases smoothly, R drops down strongly. The difference $G_I - R$ is the kinetic term $\Delta E_{kin}/\Delta A$ which accelerates the crack propagation. It has been measured that a crack is accelerated to about 1/3 of the transverse sound velocity [9.4]. The crack propagation becomes unstable. In fast moving crack tip zones, however, there grow up processes, which increase R and decelerate the crack. When finally R is as low as G_I then crack arrest occurs. For a further propagation step a new load step is necessary [9.5].

FRACTURE BEHAVIOR OF POLYMERS

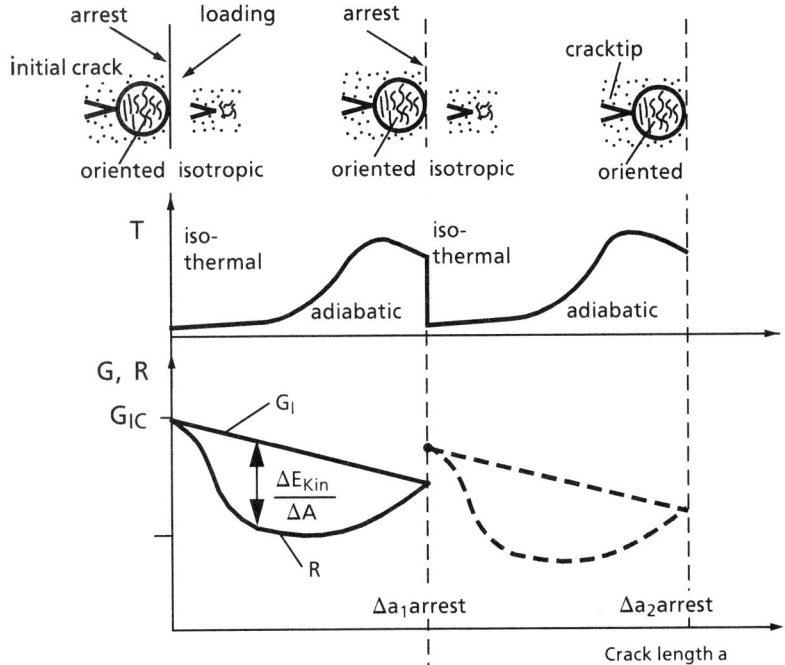

Fig. 9.4. Schematic plot of G_I, R and T as a function of the crack propagation length a during unstable crack propagation steps.

There are two main questions:

- Why does R increase during unstable crack propagation?
- What is the reason for a strong drop of R?

The first question can be answered by adiabatic heating of the fast running crack tip. Friction, chain scissions and deformations generate heat, which causes a temperature rise ΔT at the crack tip. At low temperatures the rise is especially high since the specific heat is low. For fast running cracks the amount of heat removed by thermal conductivity is negligible and adiabatic conditions exist at the crack tip. The adiabatic temperature rise ΔT_{ad} is rather high, as shown in Section 9.4.2. As a consequence, the size of the plastification zone at the crack tip is enlarged which increases the energy of plastification energy E_{pl} and thus R. In the heated and strongly loaded crack tip zone some alignment of polymer chains occurs which additionally increases R. If R just exceeds G_I, the crack arrests after a certain crack propagation length Δa. The plastic zone gets frozen under load and is printed as an arrest line in the fracture surfaces.[1] The broadness of an arrest line can be used as a measure of the plastic zone size.

The second question is implicated in the answer to the first one. It can be assumed that R is increased because of some alignment and the large plastic

---

[1] The fracture surfaces become visible, when the specimen is completely broken after the measurements.

zone. Thus, the initiation of a new crack step by external tensile loading starts with an increased R. The crack however runs isothermally from partially oriented into unoriented, non plastified bulk material. Therefore, R becomes smaller and the crack is accelerated till selfheating, plastification and small alignment at the crack tip again stop the crack. A schematic plot of G_I, R and the temperature T during crack propagation steps as a function of the crack propagation length a is shown in Figure 9.4. The process of crack arrest of polymers is thus an interplay of temperature rise → plastification → small orientation → crack arrest → freezing of orientation. This model is supported by fracture processes treated in Section 9.4.

9.2.4 Fracture Models

The stress concentration at crack tips under load can be calculated by fracture mechanics. Various models are used and two of them are briefly repeated. The Irwin model describes cylindrically shaped plastic zones at the crack tip while the Dugdale model is more adapted for the wedge shaped tips at crazes. The basic relation for calculating the stress distribution in the vicinity of a crack tip is given by [9.6]

$$\sigma_{ij} = \frac{K_I}{\sqrt{2\pi r}} f_{ij}(\theta, z) \tag{9.4a}$$

Cylinder coordinates r, θ, z are used with a coordinate system fixed at the crack tip. $f_{ij}(\theta,z)$ is a geometrical function. For θ=0 (crack plane) it holds

$$\sigma_{ij} = \frac{K_I}{\sqrt{2\pi r}} \tag{9.4b}$$

Depending on the model, a finite crack tip zone is assumed which avoids a singularity at r=0.

Irwin Model. The stress distribution in the crack plane (θ=0) of an elastic body due to Eq.9.4b is shown by the hatched curve in Figure 9.5. Since fracture strength is limited by a material specific yield strength σ_F no singularity arises at r=0. Instead, a plastic zone is built which changes the stress distribution after plastic flow (solid curve). The formation of a plastic zone does not occur by uniaxial load, but by multiaxial stresses. By means of fracture criteria a reference stress σ_i can be calculated which is assumed to be limited by the yield stress σ_F. This is treated in Section 9.4.1. By using the Tresca criterion the following diameters of the plastic zones (approximately cylinders) have been derived from the Irwin model [9.6]:

$$d_I = \frac{1}{3\pi} \left(\frac{K_I}{\sigma_F}\right)^2 \quad \text{(for plane strain) }^{2)} \tag{9.5a}$$

$$d_I = \frac{1}{\pi} \left(\frac{K_I}{\sigma_F}\right)^2 \quad \text{(for plane stress) }^{3)} \tag{9.5b}$$

---

[2] plane strain condition exists in thick specimens; because of the Poisson's ratio μ a stress component along the crack tip width takes place.
[3] plane stress condition exists if no stress component along the crack tip width takes place; it is realized in relatively thin specimens.

FRACTURE BEHAVIOR OF POLYMERS

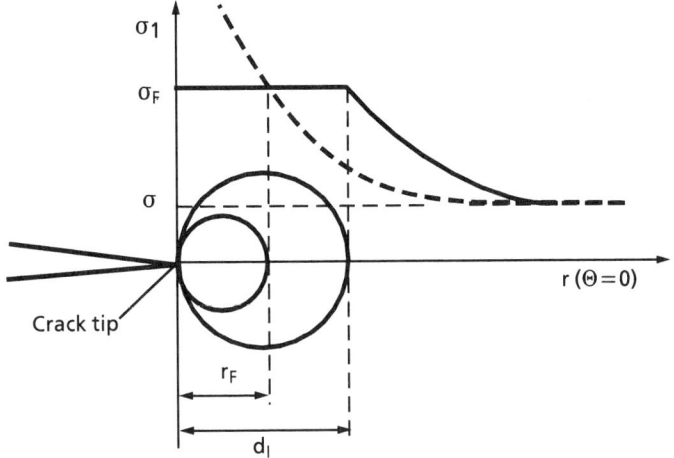

Fig. 9.5. Stress distribution and plastic zone at the crack tip (Irwin model).

The diameter of the plastic zone depends on the stress intensity factor K_I and the yield stress σ_F. Both are functions of temperature and the deformation rate. The plastic zone diameters get larger at increased temperatures. Typical ranges of values are:

$$d_I \approx 0.03 \text{ to } 0.4 \text{mm at } 4.2K$$
$$d_I \approx 0.1 \text{ to } 2 \text{mm at } 77K.$$

This is roughly consistent with experimental results which are gained from the thickness of arrest lines mentioned in the preceeding section. It can be used for calculating the plastification volume and its heat content (see Section 9.4.1).

Dugdale Model. This model describes the stress distribution of wedge shaped plastic zones. It is assumed that during tensile loading orientation of polymer chains in the crack tip zone occurs which locally increases strength. This is described by internal compressive stresses σ_c at the crack tip zone in this model. A sketch of the stress situation is shown in Figure 9.6. The solution of this problem is made in such a way that no singularities occur [9.6].

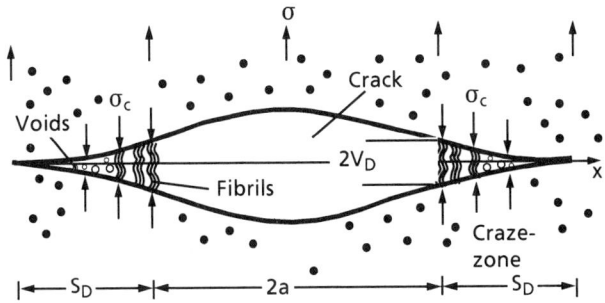

Fig. 9.6. Stress distributions of crack zones with orientation effects (Dugdale model).

The volume of the wedge shaped crack tip zone is determined by s_D and $2v_D$. For an overall stress on the specimen $\sigma \ll \sigma_F$ it holds [9.6]:

$$s_D = \frac{\pi}{8}\left(\frac{K_I}{\sigma_F}\right)^2 \qquad (9.6)$$

$$2v_D = \frac{K_I^2}{\sigma_F E^*} \qquad (9.7)$$

E^* is the reduced bulk modulus: $E^* = \begin{cases} E' & \text{(plane stress)} \\ E'/(1-\mu^2) & \text{(plane strain)} \end{cases}$

Both parameters depend on the yield strength σ_F, which controls orientation effects, especially in thermoplastic polymers. The Dugdale model is important for describing craze zones which consist of highly oriented fibrils. Chain orientations at the crack tip are a domain of the Dugdale model. At low temperatures (T<100K) where almost unstable crack propagation prevails, the formation of crazes is not well investigated. Some results of PC indicated that craze formation is enhanced by an N_2 environment even at 77K [7.7]. At RT, however, the craze zone dimensions are well estimated. Typically the craze zone length are of the order of $s_D \approx 50-200\mu m$; $2v_D \approx 2-5\mu m$ [9.6].

9.3 MEASUREMENT OF K_{IC} AND G_{IC}

The methods for measuring K_{IC} and G_{IC} are decisive, and are therefore briefly described. The important point is the choice of the right crack opening specimen. A rectangular compact tension specimen (Fig. 9.3) is not applicable for brittle materials since initial crack would lead to catastrophic failure. Chevron specimens with a triangular crack propagation area usually are applied [9.8]. A sketch is shown in Figure 9.7. The COD δ and the load P at a given rate $\dot{\delta}$ are measured.

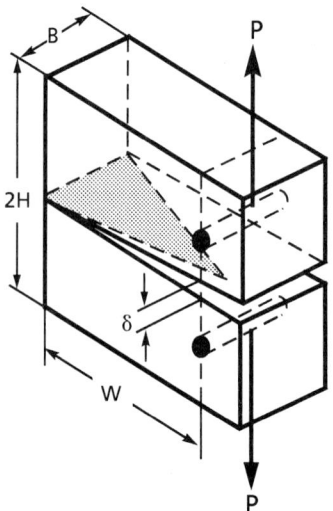

Fig. 9.7. Chevron specimen.

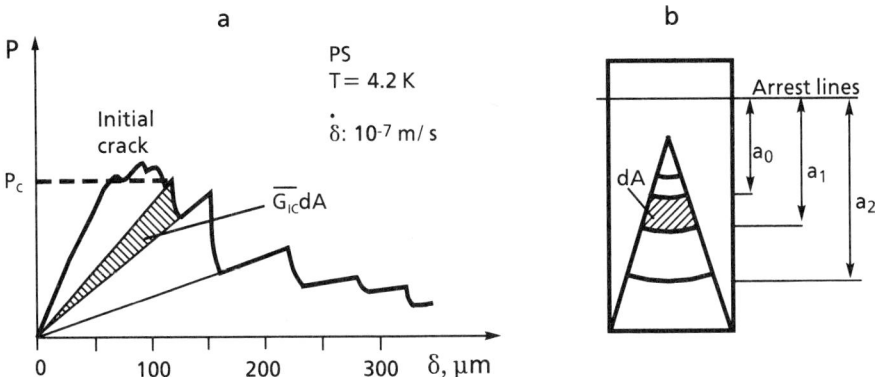

Fig. 9.8. a) Force-COD-diagrams at low temperatures b) the correlated arrest lines in a Chevron specimen.

At low temperatures crack arrest behavior prevails and a sawtooth shape of the diagram results. After loading to a critical force P_C an unstable crack propagation is initiated. By internal processes the crack stops by itself and arrest lines are printed at the crack surfaces (see Section 9.2.3). Reloading yields a new crack propagation step at a lower critical load but at a longer crack length (the values G_{IC} or K_{IC} are the same for each load step). The measurements are started after having inserted an initial crack which exhibits material specific conditions of the crack tip. When measuring G_{IC} it is important that the testing machine is very stiff and is suddenly arrested at the onset of a crack propagation; otherwise the testing machine feeds in additional energy during crack propagation and falsifies the value of G_{IC}.

Determination of G_{IC}

(1) Direct method:

G_{IC} is defined by the released energy ΔE_{el} per newly created crack area ΔA. ΔE_{el} is given by the dashed area in the diagram of Figure 9.8; ΔA is the area between correlated arrest lines. The advantage is that no correction functions are needed. The method, however, is only applicable at unstable crack propagation with crack arrest. It yields a mean value $\overline{G_{IC}}$ averaged over the propagation step Δa.

(2) Compliance method:

The method can be used at stable and unstable crack propagation. It yields a G_{IC} value at the crack initiation or at the crack arrest. The sample compliance C is measured as a function of the area A (or the correlated crack length a). It holds

$$G_{IC} = \frac{P_c^2}{2}\frac{dC}{dA} \qquad (9.8)$$

P_C is the critical force for driving or initiating crack propagation. Both methods do not necessarily yield equal values.

Determination of K_{IC}

(1) Direct method:

The critical force P_C for crack initiation is measured at a given crack length a. K_{IC} can be determined from Eq.9.1b if the correction function Y(a/W) is known. The latter can be calculated or determined experimentally by the sample compliance C as a function of the crack length a [9.8]. The combination of the critical force P_C and the given crack length a at each sawtooth of Figure 9.8 yields the same value of K_{IC}. This method is applicable for stable and unstable crack propagation.

(2) Determination from G_{IC}: It is used the general relation for stable crack propagation
$$K_{IC}^2 = E^* G_{IC} \tag{9.9}$$

E* is the reduced modulus at plane strain or plain stress conditions (see Eq. 9.7).

Dependence on the Preloading Time

An experimental result is worth mentioning since it supports the assumption of orientation effects at a crack tip under tensile load. The force-COD diagrams of Figure 9.8 usually are measured for reloading immediately after each crack propagation step. If, however, the load P_R is kept over a sufficient waiting time Δt, the reloading curve has changed. The critical force P_2 is higher than P_1 and with it the value of K_{IC}. The force-time diagrams are plotted schematically in Figure 9.9.

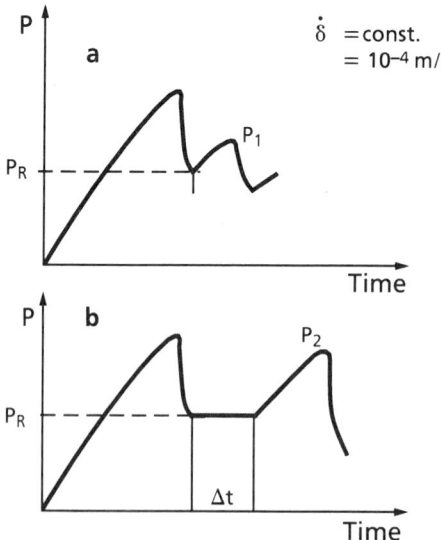

Fig. 9.9. Force-time diagram of unstable crack propagation
a) at immediate reloading
b) reloading after a waiting time Δt

This is a hint that orientation is increased in the crack tip zone when load at a high level works over a sufficient time period. Higher orientation increases fracture toughness K_{IC} (e.g., for HDPE at 77K the value of K_{IC} is increased by 30% at $\Delta t > 20s$).

9.4 PROCESSES PRIOR TO FRACTURE

Various types of processes occur prior to fracture, and they influence fracture behavior at different temperatures or deformation rates in a different way.

- Yielding
  craze formation
  shear bands
  necking
- Adiabatic heating.

9.4.1 Yielding

The phenomenon of large scale yielding is restricted to higher temperatures for polymers. Even the temperature T_T, characteristic of a transition from brittle to ductile fracture, usually is not much lower than RT. Only semicrystalline polymers of low molecular weight tend to lower transition temperatures (e.g., $T_T = 150K$ for LDPE [9.1]). At low temperatures (T<100K) the situation is different:

- Fracture occurs at least after a small amount only of yielding.
- No necking of tensile specimens which is characteristic of yielding by shear has been observed.
- Crazes are formed, but in a different way. At 77K crazes, at least craze-like failure lines [9.6] and at 200K branching of crazes has been observed [9.9].

More yielding is expected if adiabatic heating occurs during a fracture process. The temperature rise at a crack tip might be rather high especially at low temperatures where specific heat is small. The principles of yielding at high temperatures (e.g., at RT) are given in order to draw conclusions to their potentials at low temperatures.

Craze Formation. Crazes are observed in many polymers, mainly in amorphous thermoplastic polymers (e.g., PS, PMMA, PC...) but also in several semicrystalline and even in duroplastic polymers [9.6]. Crazes are formed at crack tips of small microcracks under tensile loading perpendicular to the crack plane. In the vicinity of crack tips (craze zones) molecular chains are oriented as fibrils in load direction. Polymer chains are pulled out of the bulk material and micronecking occurs. A schematic plot of a craze is given in Figure 9.10.

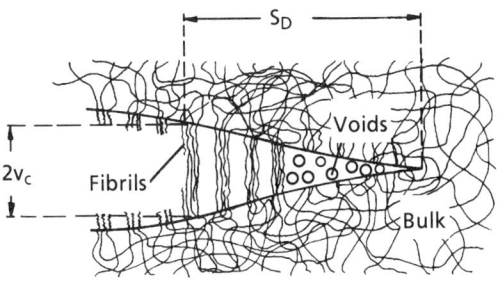

Fig. 9.10. Schematic plot of a craze.

The craze zones have a lower density than the bulk material ($\approx 60\%$). Craze zones represent a high fracture resistance since they first need high fracture energy for their formation and second they hinder crack propagation because of the oriented fibrils. Their influence is not restricted to static, but also to cyclic loading [9.6]. A general and detailed treatment is given by Kausch [9.10] and Döll [9.6]. The craze formation requires certain conditions of stress distribution. Oxborough and Bowden [9.11] proposed a criterion of largest strain in stress direction. Since craze formation increases volume, the first invariant of the stress tensor must be larger than zero. Tensile stresses generally support craze formation. As already mentioned the analytical treatment of stress distribution is given by the Dugdale model. The craze zones have a small Young's modulus but a very large fracture strain, up to about 100% at RT [9.6]. Even, if crazes are not formed at 4.2K, existing crazes will act even at low temperatures; they might be elements for achieving a higher fracture strain. The macroscopic craze density by volume is of the order of several percents in crazing polymers.

Shear Bands. The formation of shear bands is a process competitive to crazing. Both cases exhibit fibrillar structure; the preconditions, however, are different. Since crazing occurs under tensile normal stress only, shear bands are also induced by shear or compressive load. The different macroscopic fracture behavior is demonstrated on two tensile specimens in Figure 9.11. The left-hand side shows shear bands roughly under an angle of 45° which has been observed at RT, e.g., for the copolymer ABS, while the right-hand side shows crazing of PS perpendicular to the load direction [9.12; 9.13]. A craze-like pattern has also been observed on tensile specimens at 77K.

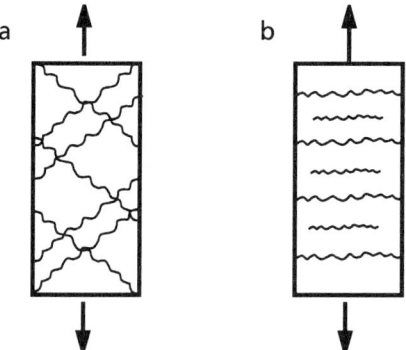

Fig. 9.11. Yielding of tensile specimens. a) by shear bands b) by crazing

The dominance of either processes depends on the type of polymer and the experimental conditions (temperature, load rate). From fracture analysis it is known that uniaxial tensile stress can be substituted by shear stress at 45°. Originally this idea has been developed for crystalline solids which are pre-destinated for shear slipping. For polymers with an entangled chain structure this process is less obvious. For cross-linked polymers shear slipping is suppressed and even for thermoplastic polymers there might be induced other fracture processes (e.g., crazing) prior to yielding. Some predictions about the dominance of a fracture process can be made by yield criteria [9.13]. For a multi-axial load system (e.g., tension, shear) there is assumed that the principal stresses combine to a reference stress, which can be compared with a critical

FRACTURE BEHAVIOR OF POLYMERS

value $\sigma_F(T,\dot{\varepsilon})$. The principle of combinations and the nature of σ_F are the topic of the different yield (or fracture) criteria.[4] The critical parameter $\sigma_F(T,\dot{\varepsilon})$ generally is a function of temperature and of the deformation rate. As a general trend the prediction holds that under tensile load crazing occurs prior to shear yielding (for foils the situation is different [9.14]).

Necking. Most semicrystalline polymers show the phenomenon of necking under tensile load at elevated temperatures. Even if this macroscopic behavior looks similar to that of metals, the microscopic processes are different for semicrystalline polymers. At the necking zones the crystallites are reoriented and aligned. At further elongations the crystallites and tie molecules are more and more oriented. This process of cold drawing needs no external heat but generates internal heat. The horizontal yielding line found at those stress-strain diagrams is misleading, and is due to the fact that nominal stresses and strains are used which are related to the initial sample dimensions. The reduced cross-sectional area of the strain hardened material between the necking zones has, of course, a higher strength. At low temperatures (for most polymers already below 200K) chain scission and brittle fracture occurs prior to large scale yielding in most cases.

9.4.2 Adiabatic Heating

Local heating at a crack tip is induced by friction, chain scission or other dissipative processes when load is applied. This effect is especially strong at low temperatures since specific heat is small, and even low heat pulses rise temperature drastically. Adiabatic conditions exist at unstable crack propagation where the rate for heat generation is lower than for its removal by thermal conductivity. The chain scission, e.g., during crack propagation, yields similar heat pulses at any temperatures but at low temperatures a much larger temperature rise results. The temperature rise can be calculated by the general relation

$$\frac{\partial T}{\partial t} = \frac{\partial}{\partial t}\left(\frac{Q_{diss}}{\rho c}\right) - a_T \, \text{divgrad} \, T \qquad (9.10)$$

with $a_T = \kappa/(\rho c)$: thermal diffusivity; κ: thermal conductivity; ρ: density; c specific heat; Q_{diss}: dissipated heat at the crack tip.

---

[4] Two criteria are mentioned [9.15]:

Tresca criterion: $\qquad \sigma_{max} - \sigma_{min} = 2\tau_y = \sigma_F$

It combines the maximum and minimum principal stresses with the shear yield stress τ_y.

von Mises criterion:

$$\sigma_i = \frac{1}{\sqrt{2}}\left[(\sigma_1-\sigma_2)^2 + (\sigma_2-\sigma_3)^2 + (\sigma_3-\sigma_1)^2\right]^{1/2} = \sigma_F$$

The maximum energy to be stored in a body before yielding occurs is expressed by the principal stresses σ_1, σ_2 and σ_3. The principal stresses are obtained by transformations to a coordinate system where shear stresses vanish. Principal stresses are normal stresses, their numerical values, however, are combinations of the applied tensile and shear stresses.

The coordinate system is assumed to move with the crack tip. [5] The second term of Eq.(9.10) takes care of heat removal. At fully adiabatic conditions

$$\Delta T_{ad} = \frac{Q_{diss}}{\rho\, c(T)} \qquad (9.12)$$

In most cases it is difficult to measure ΔT directly in the crack vicinity [9.3]. The calculation even by Eq.(9.12) is difficult since Q_{diss} is difficult to estimate. Before doing this it must be shure that an adiabatic state exists for fast running cracks, and that Eq.(9.12) is applicable. To this aim two characteristic times are compared:

- Thermal relaxation time τ_t which describes the time of heat removal.
- Heat generation time t_w.

The comparison of τ_t and t_w yields a criterion for isothermal or adiabatic conditions:

$\tau_t \ll t_w$ isothermal crack propagation
$\tau_t \gg t_w$ adiabatic crack propagation
$\tau_t \approx t_w$ temperature rise influenced by thermal diffusivity.

A long cylinder of small diameter d is used for the plastic zone of a crack tip. In the middle axis a temperature rise ΔT is assumed to be generated by dissipative fracture processes. Thermal diffusion along the radius d/2 of the plastic zone lowers ΔT exponentially with time. The thermal relaxation time τ_t is the time needed for decreasing ΔT to $\Delta T/e$. For a cylinder it holds (see Eq. 5.30)

$$\tau_t \approx 1.1 \frac{d^2}{4 a_T} \qquad (9.13)$$

Typical values of the thermal diffusion coefficient a_T are given in Figure 5.25. The heat generation time t_w is given by the crack velocity $\dot{a}$ and the plastic zone diameter d.

$$t_w \approx \frac{d}{\dot{a}} \qquad (9.14)$$

The maximum of $\dot{a}$ during unstable crack propagation has been measured to be 1/3 of the transverse sound velocity of the polymer. Calculations show that for unstable crack propagation it holds as an order of magnitude at low temperatures

$$\tau_t/t_w \approx 10 \text{ to } 500$$

That means that fully adiabatic conditions exist for unstable crack propagation and Eq.(9.12) can be used for calculating the adiabatic temperature rise.[6]

---

[5] A different representation is given in a fixed coordinate system: The temperature rise ΔT at a time t and at a distance r from the crack tip can be calculated by [9.16]:

$$\Delta T(t, r) = \frac{Q_{diss}}{c\rho\,(4\pi a_T t)^{1/2}} \exp\left(-\frac{r^2}{4 a_T t}\right) \qquad (9.11)$$

[6] Example for HDPE at an environmental temperature of 77K: $\dot{a} \approx 10^3$ m/s (maximum value); $d \approx 0.5$mm; $a_T(77K) = 5\cdot 10^{-4}$ m^2/s; a mean value $a_T(110K) \approx 7\cdot 10^{-4}$ m^2/s is used when adiabatic temperature rise is included. More exact estimations are done by iterations.

FRACTURE BEHAVIOR OF POLYMERS

Estimation of the dissipated energy density Q_{diss} is difficult. What can be measured is the released energy per crack area, G_{IC}. This can be used for estimating the upper limit of energy release due to the plastic zone area of diameter d. Assuming as an approximation a quadratic plastic zone of length d at the crack tip, the density of heat is

$$Q_{diss} = G_{IC} d/d^2 = G_{IC}/d$$

The adiabatic temperature rise is then given by

$$\Delta T_{ad} \approx f \frac{G_{IC}}{c(T) \rho} \frac{1}{d} \tag{9.15}$$

with c(T): specific heat; ρ: density. The factor f takes into account that only part of the released energy is converted into heat. A realistic value is $f \approx 0.6$ which is consistent with other experiments [9.17]. Instead of G_{IC} the correct energy rate would be R, but its value is difficult to determine. The specific heat c(T) is a strong function of T as treated in Chapter 3. Therefore Eq.(9.15) is solved by iterations. The diameter d of the plastic zone can be calculated by a model (mainly Irwin model) or determined from the broadness of arrest lines. Estimations of ΔT_{ad} for several polymers investigated, yield the following order of magnitudes [9.18]:

at 4.2K : $\Delta T_{ad} \approx 40K - 70K$

at 77K : $\Delta T_{ad} \approx 50K - 100K$

9.5 DEPENDENCE ON DEFORMATION RATE AND TEMPERATURE

9.5.1 Tensile Fracture Stress and Strain

As an example, for polycarbonate the stress-strain diagrams at different strain rates $\dot{\varepsilon}$ are plotted in Figure 9.12a. Especially the fracture values depend strongly on $\dot{\varepsilon}$. The dashed line marks the fracture line. It reflects the fact that ε_{UT} surpasses a minimum. The reason for a minimum of ε_{UT} and a rise of the fracture properties is an interplay of different processes.

Fracture stresses and strains at 77K versus strain rate $\dot{\varepsilon}$ are plotted in Figures 9.12b,c for several polymers. The strain values are normalized to their quasi-static values ε_{UTO}. At low strain rates there is no influence; at increased strain rates polymers become more brittle and fracture strain decreases. The astonishing feature is, however, that the fracture stress and strain increase at high strain rates. The polymers become more tough and ductile. What is the reason of this strange behavior? The answer anticipated, again is adiabatic heating. The processes at various strain rates are discussed on a simple model [9.19]:

Each polymer has microcracks, and one will be the nucleus of macroscopic fracture. Different processes at a crack tip occur at different strain rates. The fracture processes are controlled mainly by two scales of time: mechanical relaxation time τ and thermal relaxation time τ_t which controlls the heat removal. Their relative influences are determined by the strain rate. The dominant processes will be discussed at different ranges of strain rate (or loading time to failure). The loading time is compared to τ and τ_t.

Low strain rate (loading time $\gg \tau$):
  small isothermal, plastic deformations at the crack tip increase the crack resistance and make the fracture strain higher.
Medium strain rate (loading time $<\tau$):
  time is not sufficient for plastic deformations at the crack tip; polymer becomes more brittle and fracture strain is decreased.
High strain rate (loading time $>$ thermal relaxation time τ_t):
  increased heat power generation at the crack tip by deformation and friction, but most heat is removed by thermal conductivity.
Very high strain rate (loading time $<\tau_t$):
  heat generation is faster than its removal; quasi-adiabatic heating at the crack tip occurs which enhances plastic deformations; fracture stress and strain are increased.

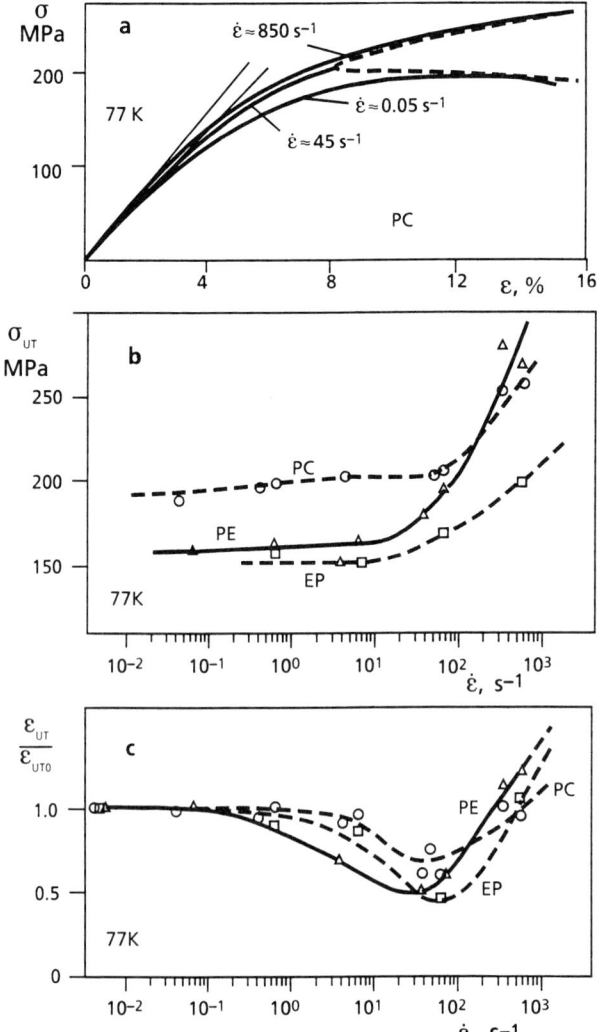

Fig. 9.12. Fracture stress and strain at 77K.
a) stress-strain diagrams of PC at different strain rates $\dot\varepsilon$.

b) ultimate tensile stress versus $\dot\varepsilon$.

c) ultimate tensile strain versus $\dot\varepsilon$ (the strain values are normalized to the static values).

Adiabatic heating is pronounced at low temperatures, where specific heat is small. Similar results, however, at RT have been found by Kausch on PEEK [9.10]. The maximum strain rate considered in Figure 9.12 is not high enough for a fully adiabatic condition. The temperature rise at the crack tip and thus the fracture properties are still a function of $\dot{\varepsilon}$. They probably could be increased further. The temperature rise at the crack tip is of the order of $\Delta T_{ad} \approx 50K$ to 80K at 77K. It can be assumed that secondary glass transitions, at least their low-temperature tails, enhance plastification. For the polymers considered, the sequence of strain rates, where fracture properties start rising, is a hint for this. HDPE has the lowest secondary glass transition temperature ($\approx 155K$) and its increased crack tip temperature comes closer to the dispersion region than for the other polymers considered. In other words, for HDPE a sufficient temperature rise is already achieved at a relatively lower strain rate. Epoxy resins have a higher secondary glass transition temperature ($\approx 230K$) and need a higher strain rate for achieving a sufficient temperature rise.

Fracture processes with adiabatic heating are possible also at very low temperatures, e.g., at 4.2K. The temperature rise, however, is smaller and not sufficient to reach a dispersion region. The specific heat is smaller, but at 4.2K the thermal diffusivity is much larger and causes more heat removal. Some compensation could be archieved with a very high strain rate, but there are limitations; ca. 1/3 of the sound velocity is the maximum speed for load transfer.

9.5.2 Critical Stress Intensity Factor

The rate dependence of the stress intensity factor is determined by the fracture processes involved. Stable, isothermal and unstable, adiabatic crack propagations exhibit a different behavior. The correlation of K_{IC} and the fracture stress σ_c of Eq.9.1d can only be applied if the same fracture processes determine both, K_{IC} and σ_c. Polymers with the same static tensile strength do not necessarily have the same K_{IC} and vice versa. The reason is indirectly due to the dependence on the deformation rate. Static strength is measured isothermally at low load rates where no heating and not much plastic deformation at the crack tip occur at low temperatures. K_{IC}, however, is measured in a different way. In most cases crack opening specimens are applied which are furnished with a material specific initial crack; (otherwise, K_{IC} would depend on the quality of notch machining). At low temperatures crack propagation is almost unstable, even for the initial crack. The crack tip is thus subjected to large deformations, adiabatic heating and orientation effects (see Section 9.2.3: Crack Arrest Behavior). This state is frozen after crack arrest. The measurement of the material specific K_{IC} is therefore determined by the initial crack tip, which is the result of preceding fast deformations. Consequently, a comparison between K_{IC} and the fracture stress σ_C is only meaningful if the latter is measured at a sufficient high deformation rate, which generates a similar crack tip prior to fracture. The decisive point is that the states of the (deformed) crack tips are similar when fracture stress is compared to the stress intensity factor [9.19].

An example is given for comparison of HDPE with EP. As seen from Figure 9.12b, both polymers have nearly the same static tensile strength at low deformation rates and 77K. The value K_{IC} of HDPE, however, is higher by a factor of 4.5 (see Table 9.A in the Appendix).

Real values:

$\sigma_C(PE) \approx \sigma_C(EP)$ (non adiabatic)

$K_{IC}(PE) \approx 4.5\, K_{IC}(EP)$ (adiabatic)

} no correlation

At high strain rates, Figure 9.12b shows that for HDPE the increase in fracture stress is much stronger than that of EP, and one might speculate that at very high strain rates this tendency would intensify. In this case, at a sufficient high rate $\dot{\varepsilon}$, HDPE would have a much higher value also of the fracture stress than EP.

Hypothetical speculation:

$\sigma_C(PE) \approx 4.5\, \sigma_C(EP)$ (adiabatic)

$K_{IC}(PE) \approx 4.5\, K_{IC}(EP)$ (adiabatic)

} correlation between K_{IC} and σ_C

The stress intensity factor depends generally on the following parameters:

- Mode of propagation.
- Temperature.
- COD rate.
- Environment.
- Molar mass.

The isothermal-adiabatic transition temperature which separates stable and unstable crack propagation is considerably controlled by the COD rate. Low COD rates favor stable propagation. At very low temperature only unstable crack propagation has been abserved. No strain rate (COD rate) dependence of K_{IC} is expected since it is measured at the initial crack, whose tip is formed by unstable crack propagation and is thus "adiabatically pretreated." This expectation is correct as shown in Figure 9.13a. At 4.2K no dependence on the COD rate has been observed within three orders of magnitude. This also true for PMMA at 77K as seen from Figure 9.13b.

The temperature dependence of K_{IC} has been a matter of controverse discussions. Several authers expected that K_{IC} at stable crack propagation is influenced by secondary glass transitions [9.7]. The results on HDPE in Fig. 9.14 exhibit a rather temperature-independent behavior even in the vicinity of the glass transition at 160K.

An influence on environment has been found especially for polycarbonate (PC). The mechanical properties are different at a nitrogen-, helium- or argon environment. Its influence, however, seems to be less sensitive to K_{IC} than to G_{IC}. The following K_{IC} values have been found for PC at a COD rate of 10^{-6}m/s and 77K:

$K_{IC} \approx 4.7$ MPa m$^{1/2}$ at gaseous helium atmosphere;
$K_{IC} \approx 5.2$ MPa m$^{1/2}$ at LN$_2$ environment.

The molar mass is a parameter which strongly influences the K_{IC} values. For example, the K_{IC} values of polyethylene, namely HDPE, MDPE or LDPE, differ within a factor of ten at RT[7] [9.22].

---

[7] Maybe, it's more than a joke that K_{IC} is a very sensitive parameter for measuring the molar mass.

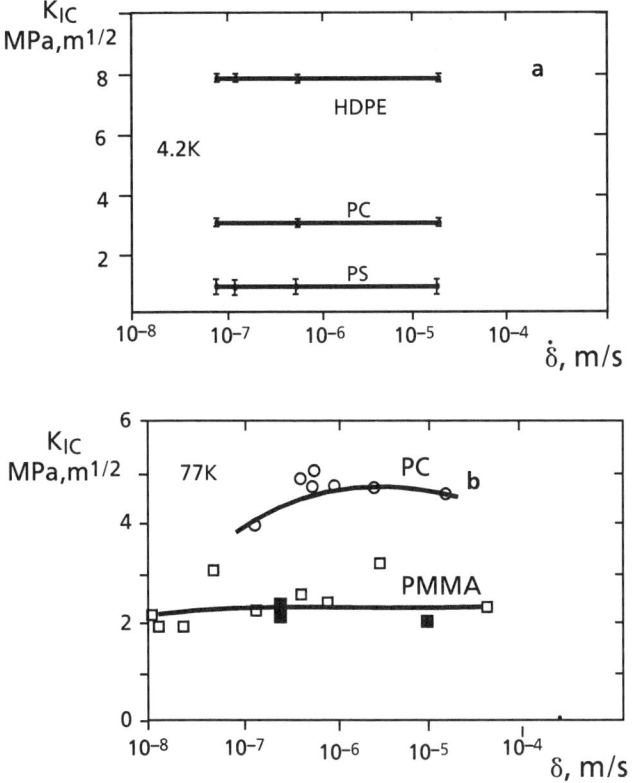

Fig. 9.13. Critical stress intensity factor versus COD rate a) at 4.2K; b) at 77K.

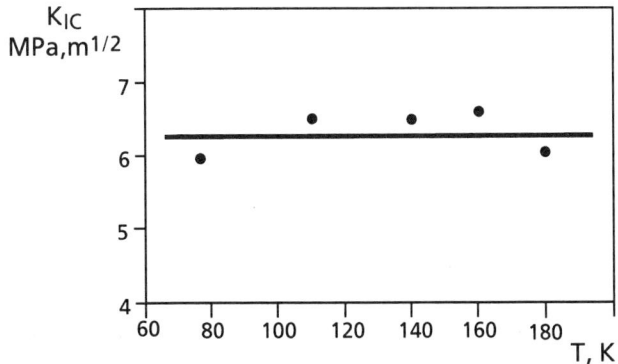

Fig. 9.14. Critical stress intensity factor K_{IC} of HDPE versus temperature for stable crack propagation at a COD rate of $7 \cdot 10^{-6}$ m/s.

9.4.3 Critical Energy Release Rate

The dynamics of crack propagation is described by the critical energy release rate G_{IC}. The question arises about the dependence on the strain rate of G_{IC} and how it influences the propagation mode. It might be expected that at extremely low COD rates stable crack propagation will be favored. This was the reason for testing at very low COD rates. The energy release rate depends on the type of polymer and on the parameters mentioned for K_{IC}. In addition, the crack velocity is a decisive parameter. Several correlations have been established between G_{IC} and the crack velocity [9.22]. The situation is rather complex and very material-dependent, except at very low temperatures. At 4.2K only unstable crack propagation has been found within three orders of magnitude.

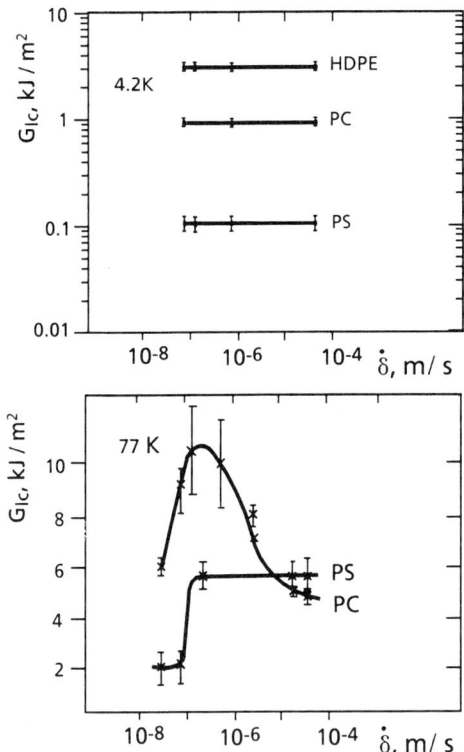

Fig. 9.15. Critical energy release rate versus COD rate at 77K. **a)** at 4.2K; **b)** at 77K.

At 77K the situation is more complex. As seen from Fig.9.15b, only at extremely low COD rates ($<10^{-7}$m/s) stable crack propagation occurs, e.g., for PS[8]. PC shows only unstable crack propagation and in LN_2 environment a strong rate dependence of G_{IC} and high values (up to 10kJ/m^2). In a helium gas atmosphere the values are on the order of $G_{IC} \approx 1.5$kJ/m^2 at 77K and it is rather independent on the COD rates. For PMMA only unstable propagation and only a small rate dependence have been found at 77K.

---

[8] Probably creep deformation is involved in this very slow crack propagation.

FRACTURE BEHAVIOR OF POLYMERS

The energy release rate both at stable and unstable crack propagation is depending on temperature. An example of stable propagation is given for HDPE in Fig.9.16.

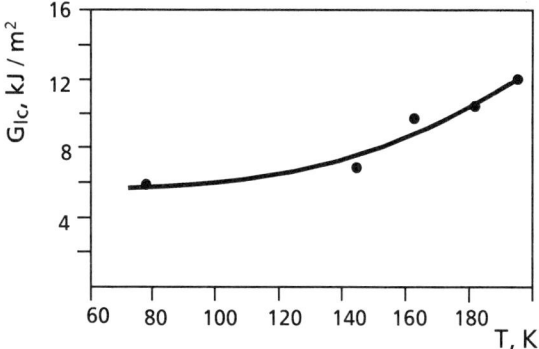

Fig. 9.16. Temperature dependence of G_{IC} for HDPE at a COD rate of $7 \cdot 10^{-6}$ m/s.

Several authors found some influence of secondary glass transitions on G_{IC} (see Ref [9.22]). The plastification energy is increased by a glass transition, but simultaneously the modulus drops. Some data are compiled in Appendix 9A.

9.6 FATIGUE STRENGTH

At cyclic loading damage is accumulated till fracture occurs. The fatigue behavior depends on the load (stress or strain) amplitude and the mean load. The load levels usually are characterized by the ratio

$$R = \frac{\text{minimum load}}{\text{maximum load}}$$

Two important types of uniaxial loading are:

- tensile threshold : $R = 0$ [9]
- alternating tension-compression : $R = -1$

Examples of the load-time profiles are given in Figure 9.17. For threshold loading the stress amplitude σ_a is superimposed to a constant stress σ_m.

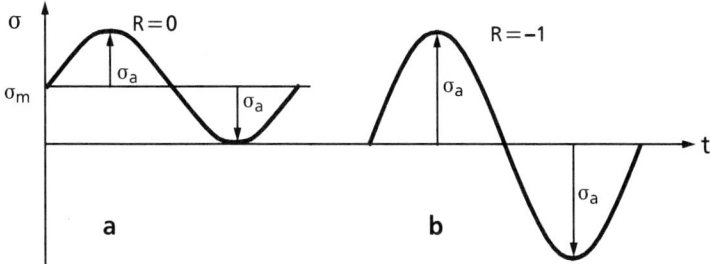

Fig. 9.17. Load-time profiles a) tensile threshold; b) alternating tension-compression.

---

[9] Usually for tensile threshold loading a finite value $R = 0.1$ is applied because of experimental reasons.

At a given load maximum the amplitudes σ_a for R=−1 are twice that for R=0.1. The load amplitudes usually are plotted versus the load cycles N to failure in so-called Wöhler diagrams (SN-diagrams). Several examples of uniaxial- and shear fatigue cycling are shown in Figures 9.18 to 9.21. During fatigue loading the strength degrades, but not the moduli. At least at low temperatures the moduli are nearly not changed by fatigue cycling. The dependences on temperature and frequency at a ratio R are the important questions. From the material point of view no significant frequency dependence is expected at low temperatures. A frequency dependence, however, arises from internal heating by friction during fatigue cycling.

9.6. Self-heating by Fatigue Cycling

The temperature rise by internal heating decreases the fatigue life. The volume of the specimen is heated by friction while heat transfer occurs at the surface. In the middle of the specimen there is a temperature rise, which depends on the sample geometry and the cooling environment. The dissipated power W_{diss} per volume is roughly given by [9.21]

$$W_{diss} \approx \pi \tan\delta \, E' \, f \, \varepsilon_a^2 = \pi \tan\delta \, f \, \frac{1}{E'} \sigma_a^2 \; ; \; W/m^3 \tag{9.16}$$

ε_a or σ_a: load amplitudes; f: frequency; E': modulus. The loss-factor $\tan\delta$ describes heating at load amplitudes in the linear viscoelastic range. In the nonlinear viscoelastic range the dissipated power is higher and can be obtained from hysteresis measurements (see Section 7.8).

Without cooling the temperature rise is determined by the specific heat c which is very low at 4.2K. The decisive parameter is the ratio $\tan\delta/c$, which for most polymers is 20 to 50 times higher at 4.2K than at 77K. With external cooling the decisive parameter is the thermal diffusivity $a_T = \kappa/(c\rho)$. It determines the time for reaching the steady state and the temperature rise in the steady state. At 4.2K it is higher by a factor of 30 to 100 than at 77K. So the influences exerted by both parameters are roughly compensated. In the steady state the powers of heat generation H_g and heat removal H_R by heat conduction are equal

$$H_g = \int_V W_{diss} \, dV \equiv H_R = \int_A \kappa(T) \, \text{grad} \, T \, dA \tag{9.17}$$

where V is the loaded sample volume and A is the surface area, which controls heat transfer to the sample environment. A specimen strip with small thickness d and a height h and broadness b is taken. The gradient is approximated by:

$$\text{grad} \, T \approx \frac{2}{d}(T_M - T_{sur}) \equiv \frac{2}{d} \Delta T$$

T_M and T_{sur} are the temperatures at the sample middle and its surface, respectively. From Eq. 9.17

$$T_M \approx T_{sur} + \frac{W_{diss} \, d^2}{4 \, \kappa(T)} \quad \text{or} \quad \Delta T \approx \frac{W_{diss} \, d^2}{4 \, \kappa(T)} \tag{9.18}$$

κ̄ (T) is the mean thermal conductivity between T_M and T_{sur}. (This temperature difference is influenced by the thermal diffusivity). For a rod-like specimen of diameter d the temperature rise ΔT is smaller by a factor of nearly four since more surface area is exposed to the cooling liquid. Flat specimens and rods with the same thickness or radius behave similarely.

There are several advantages, which allow higher frequencies for fatigue cycling at low temperatures to be applied:

- The surface temperature T_{sur} is fixed by the cooling liquid, and high heat transfer takes place.
- W_{diss} is small since $\tan\delta$ is small.

The disadvantage is that thermal conductivity gets smaller at low temperatures. For amorphous polymers, however, the temperature dependence is little. More serious is the situation for semicrystalline polymers at very low temperatures (T<20K) where κ gets very small (see Fig. 5.10b). On the other hand $\tan\delta$ is generally smaller for semicrystalline polymers since only the amorphous domains are dissipative. At 77K the situation for semicrystalline polymers generally is better and less temperature rise takes place. Several examples will be given for a broad specimen strip with a thickness d=4mm and $\varepsilon_a=1\%$:

	at 4.2K	at 77K
HDPE :	$\Delta T \approx 0.1\,f$	$\Delta T \approx 0.2\,f$
PC :	$\Delta T \approx 0.3\,f$	$\Delta T \approx 0.4\,f$

Allowing a temperature rise of 10K at 4.2K for HDPE a frequency of f=100Hz can be applied. At 77K the temperature rise is 40K at f=100Hz for PC. This is roughly consistent with experiments. The temperature in the middle of a rod-like specimen shown in Figure 10.15 (8mm in diameter) has been measured, and the following results have been found for f=100Hz and $\varepsilon_a=1\%$:

$\Delta T = T_M - T_{sur} \approx 8K$ for HDPE at 4.2K
$\Delta T = T_M - T_{sur} \approx 30K$ for PC at 77K.

A temperature rise of 30K to 40K at 77K is not tolerable and a lower frequency has to be applied. The above results apply to $\varepsilon_a=1\%$. In most cases this is the case in the region of the fatigue endurance limit. At higher load amplitudes ε_a the situation gets worse. But there is another advantage, especially at very low temperatures. Prior to reaching the steady state, the specimen is heated continuously. This takes time and allows many load cycles before dangerous heating takes place. The velocity of heat removal is controlled by the thermal diffusivity (see Fig. 5.25). At 4.2K thermal diffusivity is high enough as to remove heat at a similar time than generated at 100Hz, e.g., for a rod-like specimen with 8mm thickness. At 77K thermal diffusivity is much smaller and only the first several hundred load cycles are safe, e.g., at $\varepsilon_a=2\%$ and 100Hz. The curves shown in Figures 9.18 and 9.19 exhibit a fatigue endurance limit, which corresponds to a strain amplitude ε_a well below 1% at R=0.1. A frequency of 100Hz is therefore tolerable.

9.6.3 Measurements

Uniaxial loading. The test machine is described in Section 10.10. Especially for compression or tension-compression tests the specimen must be stable against buckling. A specimen type which has been proven to obey this requirement is sketched in Figure 10.13. The rod-like specimen has a crossection which is increased continuously from the middle plane to the clamp region. It is tightly fixed in a split cone. Fixation occurs after cooling to 77K (or below) in order to avoid shrinkage of the polymer specimen by thermal expansion. The crossection of each fracture area is determined for calculating the correct stress value.

Shear loading. The tests are performed on thin polymer tubes which are loaded by a torsion testing machine. The main problem is the transfer of shear forces from the clamps to the gauge length. Shrinkage of the polymer tubes within the clamping jaws when cooling is another problem.

9.6.4 Results on Fatigue Cycling

Uniaxial loading. SN curves for several polymers at several temperatures and test frequencies are presented in Figures 9.18 to 9.21. It is plotted the normalized upper load σ_u/σ_{UT} versus the load cycles N. As already mentioned, for $R=-1$ the (dissipative) load amplitude is twice that for $R=0.1$.

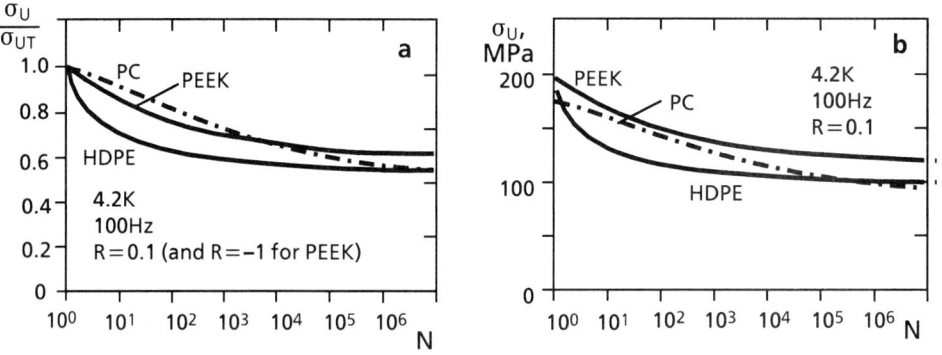

Fig. 9.18. SN curves at 4.2K a) normalized; b) tensile stress σ_U versus N.

As seen from Figures 9.18a,b the fatigue endurance limit is similar for the polymers considered. The results for PEEK indicate that at 4.2K cycling at 100Hz yields the same SN curves for tensile threshold and tension-compression loading. Self-heating, which is much higher for the latter case, does not influence the fatigue characteristic. The situation is different at 77K as shown in Figures 9.19a,b. At 100Hz the tension-compression curves show lower values, which is attributed to internal heating due to larger load amplitudes.

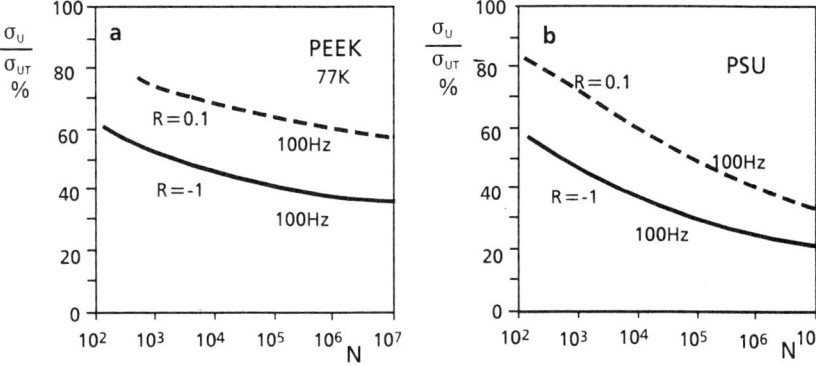

Fig. 9.19. SN curves at 77K. a) PEEK; b) PSU

In Figure 9.20 fatigue curves are plotted at 0.5Hz and 50Hz cycling. The tensile threshold curve (R=0.1) is unaffected by frequency, but not the tension-compression curves (R=-1). At 50Hz it is below and at 0.5Hz it approaches the R=0.1 curve.

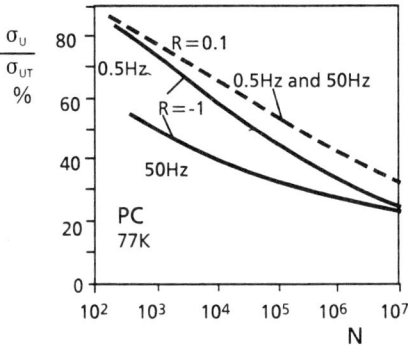

Fig. 9.20. SN curves at 77K for R=0.1 and R=-1 at 0.5Hz and 50Hz.

In Figure 9.21 the load cycles N_a to failure are plotted versus frequency for a constant stress amplitude $\sigma_a/\sigma_{UT}=0.5$ and R=-1. It can be seen that for R=-1 the load cycles N_a decrease drastically when the frequency is increased above 10Hz.

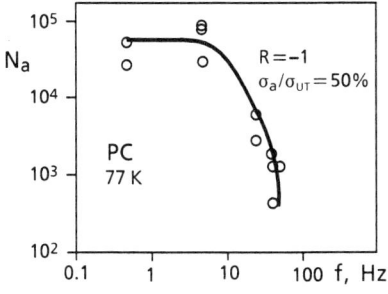

Fig. 9.21. Frequency dependence of the fatigue characteristic at a constant stress amplitude $\sigma_a/\sigma_{UT}=0.5$ for PC at 77K. N_a are the load cycles to failure.

Some examples of the temperatures dependence of the fatigue behavior are plotted in the Figures. 9.22a,b.

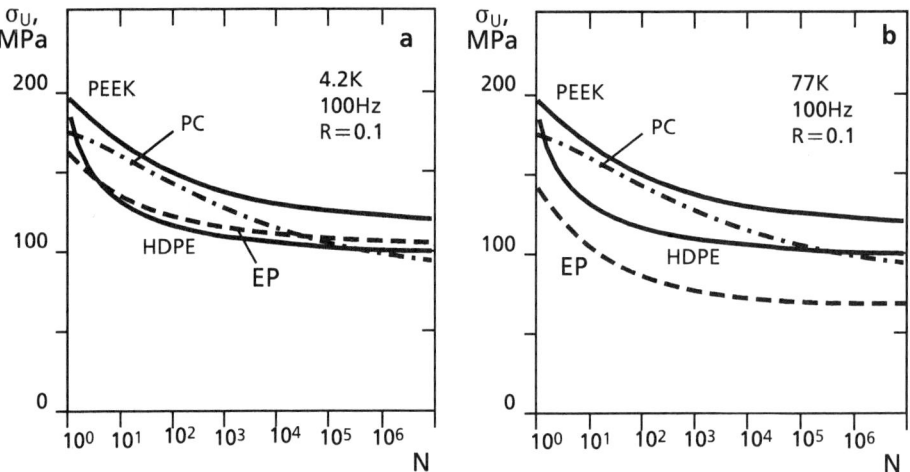

Fig. 9.22. SN curves at 4.2K and 77K of several polymers. (Epoxy resin E 162 E 113; Shell).

The poor data base available allows only guide lines to be given. A frequency dependence may arise from self-heating, which is determined by the sample size, frequency and the cooling medium. The curves above and the following statements apply to rod-like specimens 8mm in diameter.

(a) The SN curves at 4.2K are somewhat higher than at 77K.
(b) At LHe environment at least 100Hz are applicable for R=0.1 and R=-1 cycling. Both SN curves are nearly equal.
(c) At LN_2 environment up to 100Hz are applycable for R=0.1 and < 10Hz for R=-1. If the fatigue endurance limit is very low, higher frequencies can be used. The fatigue curves for R=0.1 and R=-1 are very similar at low frequencies.

Shear Fatigue Loading. Some SN curves for shear loading are compiled in Figure 9.23 for PEEK and an epoxy resin. The shear fatigue endurance limits are similar for both; PEEK, however, degrades stronger.

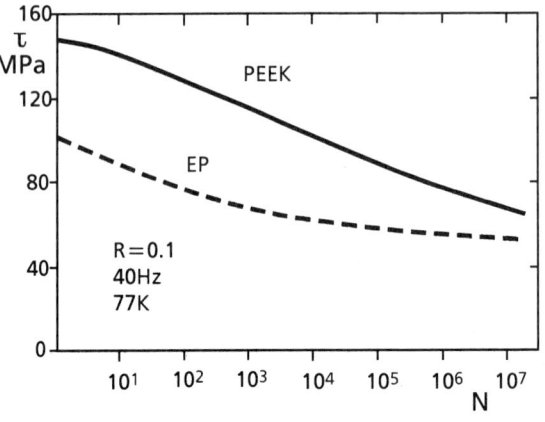

Fig. 9.23. Shear fatigue curves at 77K for PEEK and an epoxy resin (E162 E113; Shell).

9.7 Summary

(1) The mechanical strength of polymers is much lower then the strength of the polymer chains. The random entanglement of chains and stress inhomogenieties (stress concentrations) reduce the overall strength.

(2) The tensile-, compressive- and shear strengths are a strong function of temperature and deformation rate. At very high load rates adiabatic heating occurs at the crack tip which increases the fracture values.

(3) Crack propagation is described by the stress intensity factor K and the energy release rate G. At low temperatures unstable crack propagation prevails which is less rate dependent. The unstable propagation of cracks stops by itself after a certain crack length due to an interplay of released and consumed fracture energies

(4) Most fracture processes at low temperatures are influenced by adiabatic heating. The temperature rise at the crack tip is especially high since the specific heat is low.

(5) The fatigue measurements at low temperatures can be performed at high frequencies (up to 100Hz). Self-heating of the specimens constitutes the limitation. At the same upper load tension-compression cycling generates much more heat than threshold cycling. In the first case the dynamic amplitude is twice that of threshold cycling. The temperature rise in the specimen depends on its size and on the cooling environment.

(6) The fatigue behavior is only little dependent on temperature (below 100K). The fatigue characteristics are better at 4.2K then at 77K.

9.8 REFERENCES

9.1 Kinloch A. J. and R. J. Young: Fracture Behavior of Polymer; Applied Science Publishers; London-New York; (1983); p.88.
9.2 Ref. 9.1; p.93.
9.3 Schoeck, G.; Int. J. Fracture, Vol. 44 (1990); p. 1.
9.4 Kausch, H. H.: Polymer Fracture; Springer Press; Berlin-Heidelberg; (1987); p. 312.
9.5 Ref.94; p. 365.
9.6 Döll, W. in Advances in Polymer Science 52/53; Ed. H. H. Kausch; Springer Press; Berlin-Heidelberg; (1983); p. 108.
9.7 Parrish, M. and N. Brown; Nature Phys. Sci.; Vol. 237; (1972); p. 122.
9.8 Munz, D.; R. T. Bubsey and J. E. Srawley; Int. J. Fracture, Vol. 16; (1980); p. 359.
9.9 Schirrer, R. and C. Goett; J. Mat. Sci.; Vol 16; (1981); p. 2563.
9.10 Kausch, H. H. (Ed.); Advances in Polymer Science 91/92; Springer Press; Berlin Heidelberg; (1989); p. 210.
9.11 Oxborough, P. B. Bowden; Phil. Mag.; Vol. 28; (1973); p. 547.
9.12 Ref. 9.1; p. 114.
9.13 Ref. 9.1; p. 158.
9.14 Ref. 9.1; p. 138
9.15 Ref. 9.1; p. 115
9.16 Weichert, R. and K. Schönert; J. Mech Appl. Math.; Vol. 31; (1978); p. 363.
9.17 Engelter, A. and F. H. Müller; Kolloid Zeitschr. 157; (1958); p. 89.
9.18 Saatkamp, T.; Thesis, University Karlsruhe; (1991).
9.19 Hartwig, G.; B. Kneifel and K. Pöhlmann; Advances of Cryogenic Engineering; Vol. 32; Plenum Press; (1985); p. 169.

9.20 Hartwig, G. and T. Saatkamp; Advances of Cryogenic Engineering; Vol. 40b; Plenum Press; (1994); p. 1121.

9.21 Schultz, J. M.; Treatise on Materials Science and Technology; Vol. 10, Part B; Academic Press, New York; (1977); p. 601.

9.22 Ref.9.1; pp 247 and 251.

General Reading

1. Kinloch, A. J. and R. J. Young: Fracture Behavior of Polymers; Applied Science Publishers LTD (1083).
2. Kausch, H. H.: Polymer Fracture; Springer Press; Berlin Heidelberg; (1987).
3. Döll, W. and L. Könezel, in Advances in Polymer Science 91/92; (Ed. H. H. Kausch); Springer Press; Berlin Heidelberg; (1990)
4. Döll, W.; in: Advances in Polymer Science 52/53; (Ed. H. H. Kausch); Springer Press; Berlin Heidelberg; (1983).
5. Treatise on Materials Science and Technology; Vol. 10, Part B; Academic Press; New York; (1977); (Ed. J. M. Schultz).
6. Statistical Analysis of Fatigue Data; (Eds. R. E. Little and J. C. Ekvall); ASTM, STP 744; (1979).

CRYOGENIC MEASURING METHODS

Contents

10	Cryogenic measuring methods	221
10.1	Introduction	221
10.2	Cryostats	222
10.3	Temperature measurements	224
10.4	Calorimeter	226
10.5	Thermal expansion measurements	228
10.6	Dielectric measurements	230
10.7	Mechanical damping spectroscopy	232
10.8	Mechanical measurements	234
10.9	Summary	239
10.10	General references	239

10 CRYOGENIC MEASURING METHODS

10.1 INTRODUCTION

Measurements at cryogenic temperatures are more complicated since they are performed in cryostats. Low temperatures can be achieved by bath cooling with liquid helium (LHe), liquid hydrogen (LH$_2$) or liquid nitrogen (LN$_2$) at fixed temperature levels or by evaporation cooling of the sample chamber at a selected temperature which can be kept constant. The primary cooling in many cases can be done quickly by bath cooling.

For experiments which do not allow a liquid but a gaseous environment, cooling is maintained by gas convection between the cold cryostat wall and the specimen. For several experiments vacuum is required in the sample chamber after cooling down. In this case no cooling by gas convection exists. Cooling by heat removal through the sample holder in most cases is rather time consuming. A primary bath cooling and subsequent evaporation and evacuation is a faster method. If no permanent cooling through the sample holder is possible the heat flux from all connecting parts has to be suppressed by cooling elements (see the evaporator tubes in Fig.10.3).

Since specific heat of all materials is low at cryogenic temperatures, even small heat inputs can influence the temperatures of the specimens or sensors appreciably. Elements, connecting the inside and outside of a cryostat have to be optimized with respect to low heat flux and experimental requirements. Self-heating and heat input by current leads of temperature sensors depend on the experimental set-up and may also be a source of errors.

If measurements are performed at a continuously varying temperature, a temperature gradient may be established within the specimen. If a homogeneous temperature profile is required within the specimen, sufficient time is necessary for equalization of temperature. This is a restriction of the cooling or warming rate. The time scale is given by the thermal relaxation time. The thermal relaxation time τ_T depends on the geometry of the specimen and the thermal diffusivity

$$a = \frac{\kappa}{\rho c}$$

(κ: thermal conductivity; ρ: density and c: specific heat).

For a large plate of thickness d : $\tau_T \approx 1.7\, d^2/a$.
For a long rod of radius r : $\tau_T \approx 1.1\, r^2/a$.

A temperature difference ΔT between the surface and the midplane or the axis, respectively, decays to ΔT/e within τ_T. At very low temperatures (T<10K) the thermal diffusivity of polymers is rather large (see Fig. 5.15) (for a plate 1 cm thick, the value of τ_T is below one second). Above 100K the diffusivity is smaller by a factor of 100. The special problems and requirements when measuring at low temperatures are discussed in the following chapter together with the description of measuring devices. Cryogenic equipment is described for some typical apparatus.

10.2 CRYOSTATS

10.2.1 Bath Cryostat

Bath cryostats are used both for long-term experiments at the evaporation temperature of the cooling liquid and for short-term measurements during natural warming up of the cryostat. The simplest cryostat consists of two vessels, which are thermally isolated with a vacuum, a radiation shield and reflecting foils for minimizing gas convection and heat radiation, respectively. Heat losses of the inner vessel through the flange are also reduced by insulating disks and reflecting foils, which are precooled by the cold gas evaporating from the liquid cooling bath. A sketch of such a cryostat is shown in Figure 10.1.

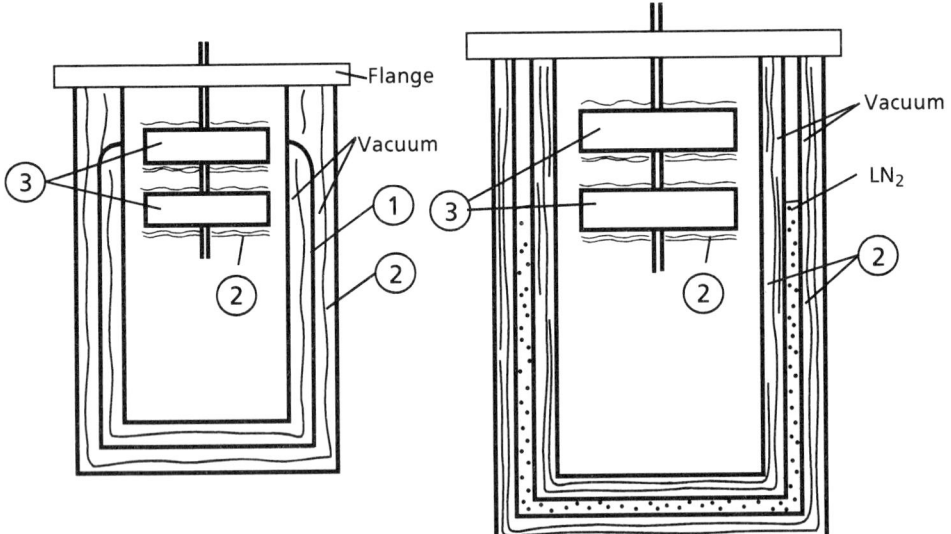

Fig. 10.1. Cryostat with (1) radiation shield (2) reflecting foils (3) insulating disks

Fig. 10.2. Helium cryostat with a jacket of LN_2. (2) reflecting foils (3) insulating disks

Improved helium cryostats are surrounded with a LN_2 jacket, which reduces LHe losses drastically. The set-up is shown in Figure 10.2. A rough estimation of the LHe losses per cryostat surface yields:

2 liter/m² per hour without a LN_2 shield;

0.5 liter/m² per hour with a LN_2 jacket.

The main cooling capacity of a liquid is given by the heat of evaporation at the boil-off temperature.

Evaporation heat of LHe : 25 J/g or   3 J/cm3;

Evaporation heat of LN$_2$: 200 J/g or 160 J/cm3.

A cooling power of 1W is achieved by evaporating 1.4 liter/hour of LHe at 4.2K. The quality of a cryogenic system is characterized by the portion of enthalpy which is utilized from the liquid and evaporated cooling medium. The cold evaporated gas can be used for precooling. For helium cryostats an optimum cooling system is important. The total enthalpy of helium from 4.2K to RT is 6.3J/mole.

A bath cryostat can also be used at temperatures above the evaporation temperature of the cooling medium for short-term experiments where an exact temperature value is not required. After extracting the cooling liquid the temperature rises slowly within the cryostat. Measurements can be performed up to RT continuously.

10.2.2 Evaporator Cryostats

The great advantage of evaporator cryostats is the possibility of regulating the temperature. A long-term temperature stabilization or a desired temperature-time profile can be achieved. Cold gas is produced by evaporation in the cryostat and fed through tubes, which are arranged around the sample chamber. A rough temperature regulation is done by regulating the evaporation rate by a heater. The exact temperature is controlled by an electrical heating system around the sample chamber. A long-term stabilization of better than 0.1K can be achieved. The principle of an evaporator cryostat is shown in Figure 10.3a.

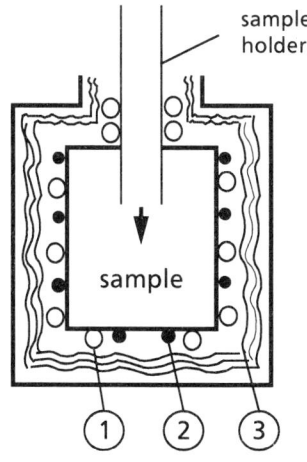
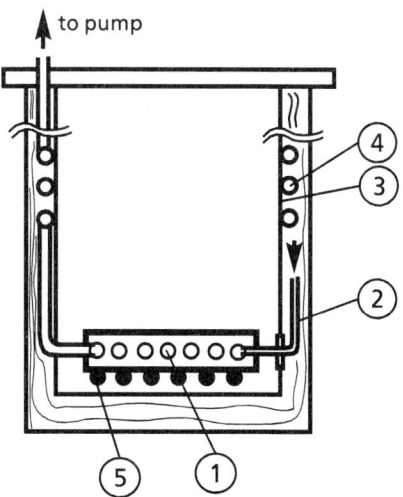

Fig. 10.3a. Principle of an evaporator cryostat
(1) evaporator tubes
(2) heating system
(3) reflecting foils

Fig. 10.3b. Evaporator cryostat
(1) cooler plate with cooling tubes
(2) capillary tube for LHe input
(3) inner wall
(4) precooling tubes
(5) heating system

A more refined evaporator cryostat makes use of the heat of evaporation near the cooler area, thus shortening the cooling time. The principle is shown in Figure 10.3b. Liquid helium is supplied through a capillary tube (2). Evaporation occurs in the capillary tube near the cooler area. The cooling element is furnished with a tubing system. The outgas of the cooler is fed through tubes for precooling of the inner wall of the chamber (3). The gas flow and the evaporation rate is driven by a pump at the outgas exit.

A combination of a bath- and evaporator cryostat is sometimes of advantage. A quick cool down is achieved by LHe and a long-term stabilisation at a desired temperature is maintained by evaporator cooling. Evaporation can be used for producing very low temperatures in the remaining liquid. By heavy pumping of LHe, subcooling below 2K is possible. If the subcooled liquid is used in an evaporator cryostat, temperatures around 2K can be achieved.

10.2.3 Special Cryostats for Short-term Measurements

Experiments at low temperatures, in most cases, require complicated cryogenic equipment. For quick measurements the situation can be simplified by the application of special cryostats. For several experiments a small inset cryostat can be adjusted to fit an apparatus which is not designed for cryogenic use. It consists of a small thermally insulating vessel, which contains the sample, the connecting elements and materials with high heat capacity. It is cooled to the desired temperature and quickly connected with the apparatus (e.g., a mechanical testing machine). Some cooling liquid in addition may be stored in the vessel. The heat capacity and the tolerable temperature rise ΔT of the specimen determine the time Δt available for measuring. An example for a special cryostat vessel is given:

- the volume of a well insulated vessel (fiber glass composite with reflecting foils) is 1 liter;
- 1 kg steel inside the pot serves as a heat capacity;
- 0.5 liter LHe is stored in the cryostat at the beginning;
- starting temperature 4.2K.

When allowing a temperature rise of $\Delta T = 5K$, a measuring time of $\Delta t \approx 30s$ is available. This is sufficient for several quick experiments. It should be mentioned that the overall measuring time with those special inset cryostats is much shorter than with ordinary cryostats whose cool down period takes several hours.

10.3 TEMPERATURE MEASUREMENTS[1]

Temperature range: $T < 0.01K$. One method of determining extremely low temperatures makes use of the temperature dependence of the magnetic susceptibility of paramagnetic salts[2]. The values, measured at somewhat higher temperatures, are extrapolated to lower temperatures. The temperature is measured indirectly by very sensitive magnetic methods.

---

[1] See G.K. White; Experimental Techniques in Low-Temperature Physics, pp. 74-126; Oxford Science Publication (1979).

[2] Adiabatic demagnetization of paramagnetic salts allows the generation of temperatures of $10^{-6}K$.

Temperature range: $T \approx 0.01K$ to $75K$. Germanium diodes in this range are very sensitive sensors. Their resistance can be optimized by doping. In the range $0.01K$ to $2K$ low-resistive, and in the range $2K$ to $75K$ high-resistive germanium diodes yield the best performances. The accuracy is better than $10^{-2}K$ at $4.2K$. One great advantage of germanium diodes is their low self-heating. At an operating current of $10\mu A$ the power generated by the germanium diodes is below $10^{-7}W$ at $4.2K$. (A heat source below $10^{-6}W$ in most cases does not influence the accuracy of temperature measurements at $4.2K$). Carbon resistors (C100) are another type of temperature sensors applied in this temperature range.

Temperature range: $T \approx 1.4K$ to $380K$. For several applications it is useful to span a large temperature range with one sensor. Silicon diodes are well-suited for this purpose. Their disadvantage at very low temperatures is their large resistance and, thus, their high heat generation. At very low temperatures germanium- or carbon sensors are superior.

Temperature range: $T > 30K$. Platinum resistors are applicable above $30K$. One great advantage is their linear dependence of resistivity on temperature. This behavior is useful when applied in electronic circuits.

In Table 10.1 some data of temperature sensors are compiled. Their temperature dependences are plotted in Figure 10.4. The data are due to an operating current of $10\mu A$. (Platinum resistors are operated at currents of 0.5 to $1mA$; the curve of Fig.10.4 is calculated for $1mA$).

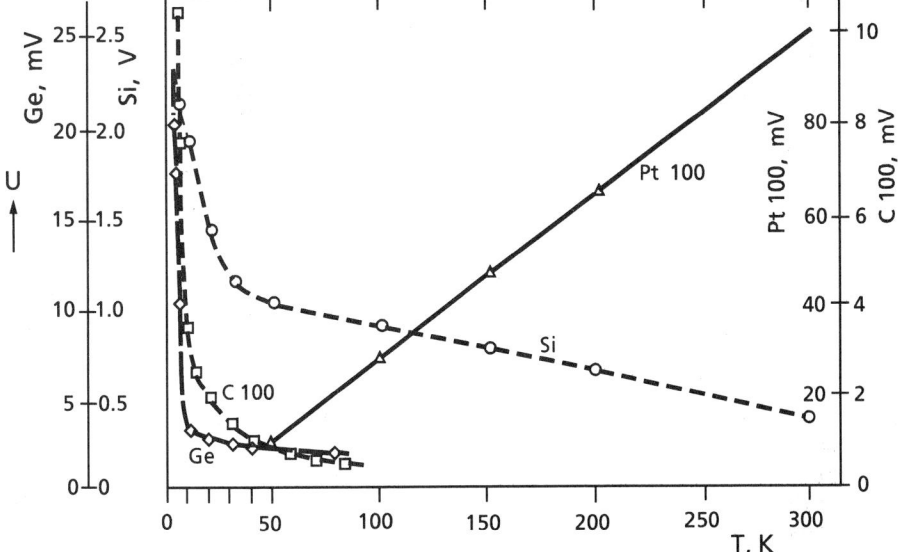

Fig. 10.4. Calibration curves of several temperature sensors versus temperature. The voltage is due to an operating current of $10 \mu A$, except for Pt 100 which is operated at $1mA$.

Each of these sensors is equipped with four leads: two for current supply and two for measuring the voltage drop at very low currents. As already mentioned, for accurate measurements it is necessary to avoid heat flux through the leads which can be done by thermal anchoring of the leads (see Fig.10.5).

Table 10.1. Resistance, sensitivity and self-heating of sensors.

Sensor	Temperature range	Resistance Ω		Sensitivity dU/dT mV/K	Self-heating at 10µA
Germanium diode					
GR 200A - 30	0.01 K – 1.5K	at 0.1K:	30	560	$1 \cdot 10^{-9}$
GR 200A - 1500	1.5 K – 75K	at 4.2K:	1600	30	$1 \cdot 10^{-7}$
Silicon diode	1.4 K – 380K	at 4.2K:	220000	50	$4 \cdot 10^{-6}$
Carbon C100	3.5 K – 75K	at 4.2K:	1000	25	$3 \cdot 10^{-7}$
Platinum Pt100	50 K – 300K	at 50K:	10	0.4	

An easy method for measuring temperatures over a wide range is through the use of thermo couples. At very low temperatures their sensitivity, however, is small.

10.4 CALORIMETER

A calorimeter is used for measuring thermal properties, such as thermal conductivity or specific heat. Two different methods will be considered for measuring thermal conductivity.

10.4.1 Rod Method

A rod-shaped specimen at opposite ends is connected to a heater and a cooling plate. The plate is cooled by evaporation of LHe. The desired temperature is controlled by the evaporation rate, and fine-control of temperature is achieved by an electrical heating system. Thermal conductivity can be determined by the steady-state method or by a well defined cooling or heating rate. For both cases an evaporator cryostat is a useful cooling method (see Fig. 10.3b). The temperatures at different positions of the specimen are measured by germanium diodes. They are connected by copper needles or thin sheets, which are glued into holes or slits of the specimen. The set-up of the apparatus is shown in Figure 10.5.

There are some main sources of error, which can be avoided by several precautions.

- Heat losses by gas convection are avoided by evacuation of the sample chamber.
- Heat radiation is reduced by a reflecting tube (10), whose temperature profile is made equal to that of the specimen by a heating circuit (11).
- Heat flux through the leads of the germanium sensors is avoided by thermally anchoring them to the tube (10). Since its temperature distribution is made equal to that of the specimen, no temperature difference exists between the connecting points (3) and the sensors (2).

By some modifications, the specific heat can be measured with this calorimeter. The specimen is not connected to a cooling plate, but fixed in the sample chamber by thin wires having a low heat input. The specimen is cooled down by cold gas convection; then the sample chamber is evacuated. Heating occurs by a resistor or by a laser pulse.

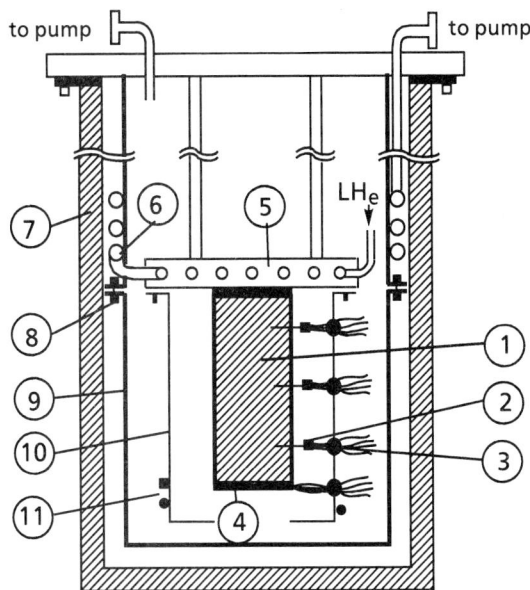

Fig. 10.5. Calorimeter for measuring thermal conductivity
(1) rod-shaped specimen
(2) germanium diodes fixed by needles
(3) thermal anchoring for leads
(4) sample heater
(5) evaporator cooling plate
(6) precooler coils
(7) removable cryostat
(8) vacuum tight flange
(9) removable vacuum chamber
(10) tube as radiation shield
(11) heater and sensor of the tube

10.4.2 Plate Method

A simple method for measuring thermal conductivity uses two disk-shaped specimens, which are sandwich-like pressed between cooler- and heater plates. The set-up is shown in Figure 10.6.

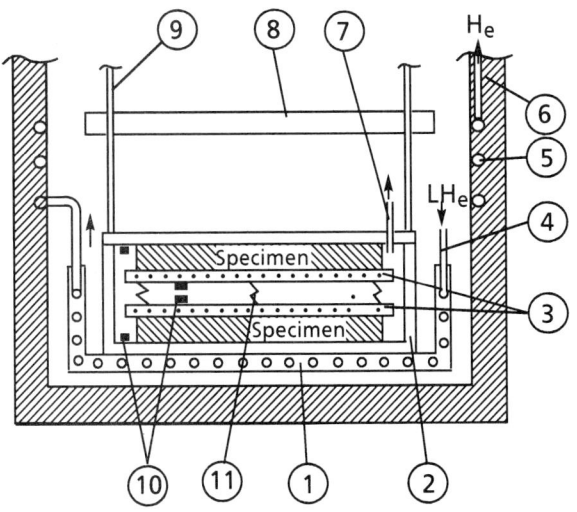

Fig. 10.6. Apparatus for measuring thermal conductivity of disk-shaped specimens.
(1) evaporator
(2) demountable cold box
(3) heater plates
(4) capillary tube
(5) outgas evaporator tubes
(6) insulating cryostat
(7) to vacuum pump
(8) insulating disk
(9) cold box holder
(10) temperature sensors
(11) springs.

The two heater plates are connected by springs (11). Usually, the two specimen disks are made out of the same material. The cooling plates are incorporated in a cold box (2), which is connected with the evaporator cooler (1). No vacuum is required outside the cold box and cooling by gas convection is possible. For sample exchange the cold box can be removed by the sample holders. From the temperatures of the cooler and heater plates (10) the temperature gradient can be determined. The measuring device is surrounded by a cold cryostat vessel, which is cooled by evaporator tubes. The great advantage of this device is the low heat loss by radiation. For thin plates the surface exposed to thermal radiation is very small. Since radiation losses increase drastically with temperature this device is useful at elevated temperatures. At low temperatures thermal boundary resistances between the plates can affect the accuracy of measurement.

10.5 THERMAL EXPANSION MEASUREMENTS

Thermal expansion can be detected electrically by capacitive or inductive methods, or optically by an interference system. Two measuring systems will be

10.5.1 Laser Interference Dilatometer

A tubular specimen is put on a mirror and covered with a semitransparent mirror. A laser beam is reflected on both mirrors. The interference pattern of both reflected beams is focussed onto a photo diode array. With temperature variations the distance of both mirrors is changed by the thermal expansion of the specimen, and the interference pattern is shifted. The shift and its direction is detected electronically by the photo diodes. The accuracy of length measurement ΔL is a quarter of the wavelength. (For red laser light and a sample length L=10cm a resolution $\Delta L/L \approx 2 \cdot 10^{-6}$ is achieved). The set-up is shown in Figure 10.7. The optical length measurement by interference methods is basically free of systematic errors. The refraction of the gaseous environment, however, can cause errors. The refractive index of a gas is temperature-dependent in the vicinity of its evaporation point. For helium gas, e.g., this is true up to 30K. Since the laser beam is influenced by the refractive index of the gaseous medium, the sample chamber must be evacuated after primary cooling. This interrupts cooling by gas convection between the sample and the cold cryostat wall. In order to avoid too rapid a temperature rise of the specimen, all connections to the specimen have to be cooled or at least thermally insulated. The sample chamber and the sample holder (4) are surrounded by tubes for evaporator cooling (9). The temperature rise is controlled by the evaporation rate or by a heater (10) which surrounds the specimen and the mirrors.

Problems may arise if impure helium gas is used. Even the residual gas in the sample chamber after evacuation might be a source of vapor deposite onto the mirrors. The interference pattern is then disturbed. This effect can be avoided by keeping the wall of the sample chamber cold enough for attracting the vapor deposite. The cylindrical electrical heater (10) around sample and mirrors is used for increasing the sample temperature.

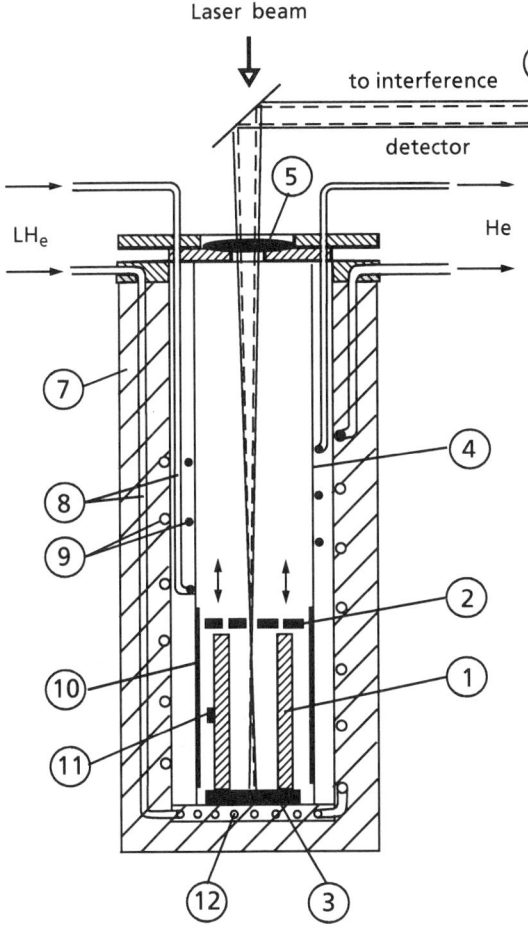

Fig. 10.7. Laser interference dilatometer
(1) tubular specimen
(2) semitransparent mirror
(3) mirror
(4) tubular sample holder
(5) focussing lens
(6) photo diode array
(7) removable evaporator cryostat
(8) capillary tubes
(9) evaporator tubes
(10) cylindrical heater
(11) temperature sensor
(12) evaporator plate.

10.5.2 Inductive Dilatometer

The dilation is measured by an inductive extensometer. The specimen is put on a glass disk, which is fixed by quartz glass rods at the cryostat flange. A push rod connects the top of the specimen with an inductive extensometer outside the cryostat. The temperature of the extensometer is electronically stabilized. The set-up is shown in Figure 10.8. A bath cryostat with LN_2 jacket is sufficient for this apparatus. The liquid helium, used for primary cooling, is boiled off, and the cryostat warms up slowly. Since no vacuum is necessary in the sample chamber, cooling exists by gas convection between the cold cryostat wall and the specimen. (For faster warmup rates, electrical heaters can be applied). The warmup rate should be low enough to allow a temperature equilibrium within the specimen. (For polymers and a specimen size 40x5x5mm^3, a rate of 1K/min should not be exceeded). The disadvantage of the push rod method is the influence of the quartz glass rods on the measured thermal expansion. Corrections have to be made for the thermal expansion of the quartz glass rods for the distance of the specimen length.

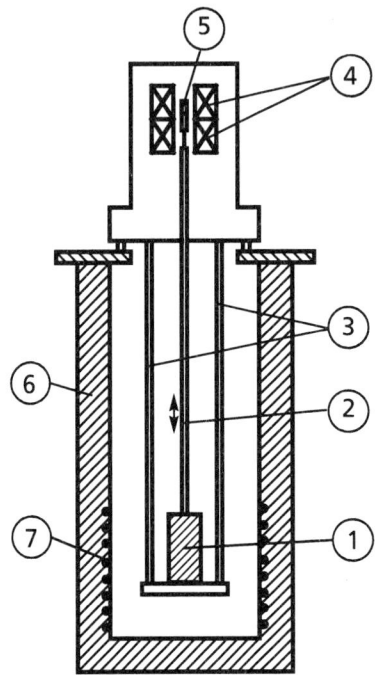

Fig. 10.8. Inductive dilatometer
(1) specimen
(2) push rod
(3) quartz glass rods (sample holder)
(4) inductive coils
(5) ferritic core
(6) bath cryostat with LN_2 jacket
(7) heater

10.6 DIELECTRIC MEASUREMENTS

Dielectric damping and permittivity can be measured up to a frequency of 10GHz. At very high frequencies microwave resonators are applied. Up to the MHz region a capacitive method is applicable. An apparatus of this kind is described in the following section.

10.6.1 AC High-Voltage Testing Apparatus

For measuring low-frequency dielectric losses and permittivity (<1kHz) or the breakdown voltage, an apparatus, as shown in Figure 10.9, can be used. A capacitor is made out of the sample material. The specimen is covered on both sides with thin metallic screens e_1, e_2 and guard rings as electrodes. The electrodes e_1 and e_2 of the thin disk area are used for measurement. The specimen is pressed by insulating springs (10) between thick electrode plates (6), (9). The measuring electrode plate (9) is connected to an electronic circuit (e.g., a Schering bridge) for measuring the capacitance. The equipment shown is designed for ca. 10kV AC voltage[3]. Application of high-voltage allows a very precise measurement of dielectric losses. An accuracy of the loss-factor tan $\delta_e \approx$ $\approx 10^{-7}$ can be achieved. A special design is required for reducing coronal discharges around the high-voltage lead and between the electrodes of the specimen. The guard ring of the specimen capacitor is therefore shaped like a Rogowski profile (see Fig. 10.10).

---

[3] Measurements in the MHz range are restricted to low-voltage AC.

CRYOGENIC MEASURING METHODS

The high-voltage lead is a tube of diameter d_1 surrounded by a grounded tube of diameter d_2 in the cryostat. Minimal coronal discharge exists for the following ratio.

$$d_2/d_1 = 2.7$$

Discharge current is minimum at very low or at high pressure of the surrounding gas. For cryogenic measurements it is advantageous to evacuate the sample chamber after cooling.

For most measurements a LHe bath cryostat with LN_2 jacket is sufficient; an evaporator cryostat, however, is of advantage. When measurements are performed during cool down, high-voltage is switched on only when the desired temperature is reached, since, otherwise polarization can be frozen in the sample.

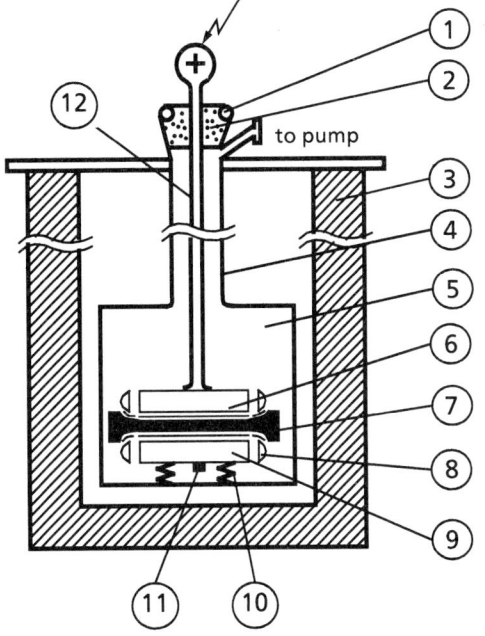

Fig. 10.9. High-voltage testing apparatus for measuring dielectric losses, permittivity or breakdown voltage.
(1) high-voltage screening
(2) insulating ring
(3) cryostat with LN_2 jacket
(4) grounded tube
(5) sample chamber
(6) high-voltage electrode plate
(7) specimen
(8) guard rings
(9) measuring electrode plate
(10) insulating springs
(11) temperature sensor
(12) high-voltage tube of diameter d_1

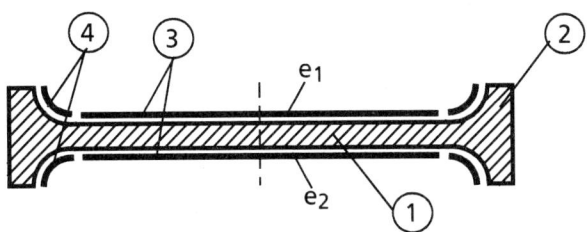

Fig. 10.10. Specimen with electrodes (Rogowski capacitor). (1) disk area (1mm thick, 90mm diameter); (2) Rogowski frame; (3) e_1, e_2 electrodes; (4) guard ring electrodes.

10.7 MECHANICAL DAMPING

Different methods are applied to cover the large frequency range which is necessary for characterizing the mechanical damping behavior:

$10^{-4} - 10^{-2}$ Hz : torsional creep

$10^{-1} - 30$ Hz : free torsional vibration (torsion pendulum)

$10^{-1} - 10$ Hz : forced oscillations (testing machine)

$50 - 10^4$ Hz : flexural vibrations (reed method)

$10^6 - 10^9$ Hz : vibrations actuated by oscillating quartz crystals.

It should be mentioned that free oscillation measurements (e.g., in a torsion pendulum) are only meaningful if damping is small. Large damping values can be better measured from the hysteresis effects with forced oscillations. In this treatment only a torsion pendulum for cryogenic use will be described. The cryogenic principles can be transposed to other such devices.

10.7.1 Torsion Pendulum

For low-frequency relaxation spectroscopy the torsion pendulum is a commonly used device. During free oscillations the torsional amplitudes decrease. A measure of the damping behavior is the ratio of successive amplitudes A_n, which is called the logarithmic decrement $\Lambda = \ln(A_1/A_2)$. The loss-factor is given by $\tan \delta = \Lambda/\pi$. The set-up of a torsion pendulum for use at cryogenic temperatures is shown in Figure 10.11a. The specimen is a rod or a strip, which is fixed by clamps (4) and connected by a thin tube (3) to an actuator (1) outside the cryostat. The actuator initiates free oscillations, which are maintained by a disk (8). The oscillations are determined electromagnetically, or optically by reflecting a light beam on a mirror (9). In the latter case the cryostat is supplied with a small window for the incoming and reflected light beam. Registration can be done photographically or by electronic methods.

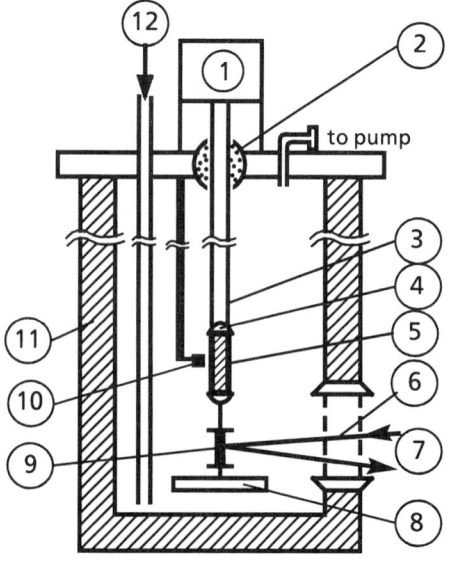

Fig. 10.11a. Simple version of a torsion pendulum.
(1) actuator
(2) vacuum tight feed-through
(3) fixation tube
(4) clamps
(5) specimen
(6) window
(7) light beam
(8) free vibrating disk
(9) mirror
(10) temperature sensor
(11) bath cryostat
(12) tube for LHe inlet

The difficulty of measuring damping at very low temperatures arises from the very low damping values. Therefore, the accuracy must be high, and even small damping contributions from the apparatus must be avoided. A first step of improving the accuracy is to accumulate the decrease of amplitudes over a large number n of oscillations

$$\Lambda = \frac{1}{n} \ln A_1 / A_{n+1} \qquad (10.3)$$

There are two main sources of undesired damping contributions from the apparatus:

- The friction in gaseous helium of the vibrating disk becomes noticeable for the small damping values found at very low temperatures. The error is about 10-20% at 4K. Pressure reduction in the sample chamber is recommended ($p \approx 0.2$ bar).
- The fixation tube (3) of the oscillating specimen may be a source of additional damping.

The feed-through (2) of this tube through the cryostat flange must be easy to rotate and tight enough to allow pressure reduction in the sample chamber. If this tube is sufficiently thick a significant damping contribution is avoided. On the other hand, the tube should be thin-walled for a low heat input. A compromise can be found experimentally by testing tubes with different stiffnesses, and measuring their additional damping contributions. The tube should be made out of invar steel which has a low heat conductivity.

The best method for avoiding these errors is to fix the actuator within the cryostat very near to the specimen. This version is shown in Figure 10.11b. There is nearly no damping from the apparatus and the vacuum chamber can easily be evacuated.

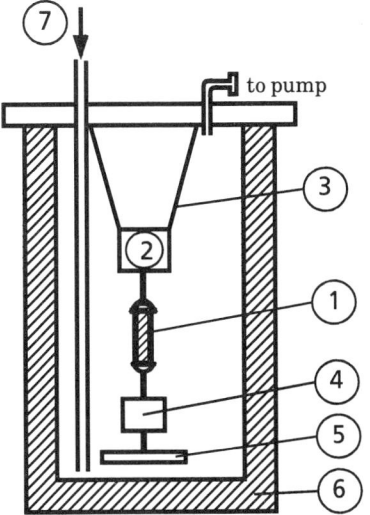

Fig. 10.11b. Torsion pendulum
(1) specimen
(2) actuator
(3) thermal insulated fixation
(4) electromagnetic oscillometer
(5) free vibrating disk
(6) bath cryostat with LN_2 jacket
(7) tube for LHe inlet

The shear modulus G' in addition can be determined from the oscillating frequency f, the momentum of inertia I of the oscillating disk and a factor F_g, which depends on the sample geometry.

$$G' = f^2 \, I \, F_g \qquad (10.4)$$

10.8 MECHANICAL TESTING

Several machines have been developed for testing the mechanical properties under tensile, compressive or shear (torsional) loads. For measuring elastic properties only low strain amplitudes are used. Special test methods are applied for different frequency ranges. As already mentioned, at low frequencies (<20Hz) the shear modulus G' can be measured by a torsion pendulum. At high frequencies ($\approx$100MHz) oscillating quartz crystals are used for generating forced longitudinal or transverse vibrations in a specimen. The moduli are determined from the wavelength of standing waves at a frequency f.

Mechanical testing machines are applied for measuring static strength and fatigue cycling at large strain- or stress amplitudes. They are restricted to low frequencies (<200Hz). The maximum frequency f of cyclic fatigue loading is limited by self-heating of specimens. The temperature in the middle of the specimen may become much higher than at the surface, and thermal stresses arise. Fatigue life is reduced, if thermal stresses are superimposed. The maximum frequency admitted at RT is of the order of 10Hz. At low temperatures much higher frequencies are tolerable depending on the sample thickness and loading mode. The reasons are very low loss-factors and a liquid cooling environment. Frequencies up to 100Hz can be applied for tensile threshold loading and a rod-like specimen[4]. A low-temperature mechanical testing machine is described in Subsection 10.8.1. A low-temperature apparatus for measuring fracture toughness and fracture energy by investigation of crack propagation is explained in Subsection 10.8.2.

---

[4] The temperature rise ΔT in the middle axis of the specimen (see Fig. 10.15) depends on the
- the heat power P, generated per volume in the specimen

$$P = \pi \tan \delta \, E' \varepsilon^2 \, f \quad ; \quad W/m^3$$

(tan δ: mechanical loss-factor; ε: strain amplitude; E': Young´s modulus).
- the heat flow D in the specimen and the heat transmission coefficient γ to the cooling environment, the specific heat c and the density ρ.

$$\Delta T \approx \frac{1}{c} (P/\rho - D \gamma).$$

Typical values of ΔT for polymers and LN_2 environment have been measured in the middle of a rod-like specimen of 8 mm diameter. The results at a frequency f=100 Hz for tensile threshold loading are:

$\Delta T \leq 40 \, K$ for $\varepsilon = 1.5 \%$

CRYOGENIC MEASURING METHODS

10.8.1 Mechanical Testing Machine

The cryogenic modification of a testing machine is characterized by two long tubes transmitting the force from the actuator to the specimen in the cryostat. The optimal length of the tubes is a compromise between low heat conduction and high stiffness. A sketch of such a testing machine is given in Figure 10.12. The large length of the actuating tube (10) is especially a disadvantage when compressive loads are put on the specimen, and buckling or a transverse shift occurs. The actuating tube (10) is therefore stabilized, by sleeve bearings (7) made out of Teflon. The machine can be modified for transmitting momenta and allow loading by torsion.

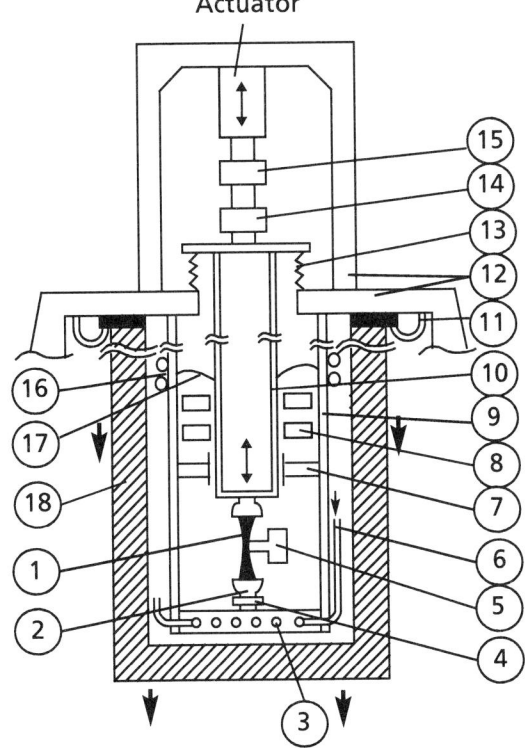

Fig. 10.12. Mechanical testing machine
(1) specimen
(2) clamps
(3) evaporator plate
(4) piezoelectric load cell
(5) extensometer
(6) capillary tube for LHe input
(7) sleeve bearings
(8) insulating disks
(9) support tube
(10) actuating tube
(11) fixation of the cryostat flange
(12) support frames
(13) bellows
(14) load cell
(15) inductive gauge
(16) LN$_2$ baffle
(17) soft thermal connection
(18) removable bath cryostat.

The cooling system consists of two elements:

- a bath cryostat (18) for fast initial cooling and
- an evaporator cooler plate (3) for maintaining long-term cooling.

The bath cryostat is advantageous for quick measurements, while the evaporator system is cheaper for long-term experiments such as fatigue measurements. When cooling is performed with LHe, the bath cryostat should be supplied with a LN$_2$ jacket for saving LHe. Evaporator cooling with LHe requires a LN$_2$ baffle (16) at the actuator- and support tube for reducing heat input. The evaporator tubes are attached to the bottom of the support tube close to the specimen. For a heavy load testing machine (>200kN) with thick-walled actuating and supporting tubes, a LN$_2$ baffle is necessary for LHe operation. Otherwise heat input disturbs LHe cooling drastically.

Stress and strain are measured by load cells and extensometers, respectively. At fast dynamic loading the load cell should be placed close to the specimen. Otherwise inertia of the heavy actuator tube induces large apparent hysteresis effects, which make load controlled measurements difficult to handle. The load cell (14) in Figure 10.12 which is placed outside the cryostat, is used for slow load detection only or for intercalibrations. A fast piezoelectric load cell (4) is arranged nearby the specimen.

In most cases strain gauges are used for the extensometers.[5] At very low temperatures special strain gauges, e.g., with a CrNi grid, are applied which show much better temperature characteristics than constantan. Their resistance is a smooth function of temperature. By applying two compensating strain gauges in a bridge circuit, the temperature dependence can be reduced further. Both strain gauges are kept at the working temperature, but only one is used for strain detection. Two examples of the temperature characteristics of compensated extensometers with CrNi- and constantan strain gauges are shown in Figure 10.13. A complete temperature compensation is not achieved since all strain gauges show little differences. Plotted is the so-called apparent strain ε_T, which is the relative resistance change $\Delta R/R$ with temperature, multiplied by the sensitivity K. It can be seen that for constantan a large temperature dependence exists below 40 K which is difficult to handle.

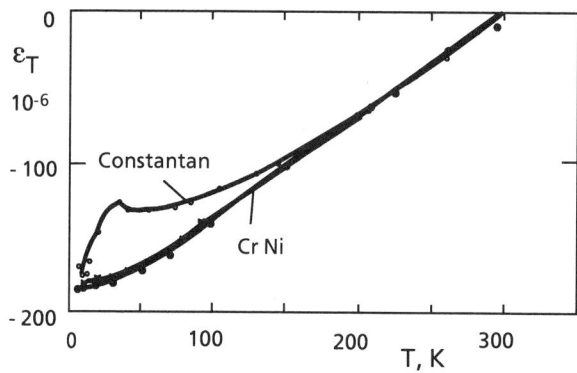

Fig. 10.13. Apparent strain $\varepsilon_T = (\Delta R/R) K$ versus temperature of two compensated extensometers supplied with CrNi- and constantan strain gauges. K is the sensitivity of the extensometer.

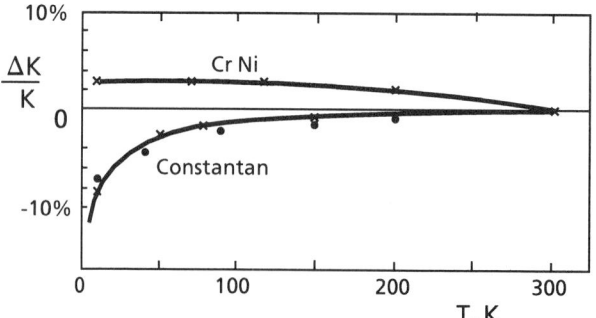

Fig. 10.14. Relative sensitivity change $\Delta K/K$ of compensated extensometers versus temperature.

---

[5] Strain gauges usually are stiff and would influence strain detection when glued onto a polymer. Therefore, soft detachable extensometers are applied at the gauge length.

The temperature dependence of the relative sensitivity K of an extensometer with CrNi- and constantan strain gauges is shown in Figure 10.14. It is a great advantage of CrNi strain gauges that sensitivity is rather independent of temperatures.

Load cells which are based on strain gauges show an analog temperature dependence as extensometers. For cyclic loading piezoelectric quartz crystals are used as fatigue resistant load transducers. Their temperature dependence is rather small.

The specimens used at cryogenic testing should be made small, or cooled down slowly in order to avoid freezing of thermal stresses. They would reduce fracture strength. An example of a rod-like specimen with a constant curvature is shown in Figure 10.15. This special type is suitable as well for tensile and compression testing. The load is induced gradually and buckling is avoided. For tension-compression fatigue testing a tight fixation is necessary. The specimen is fixed in conical clamps. Both, clamps and specimen should be precooled before fixation. Otherwise different thermal contraction will cause loosening. For the procedure of fixation, the clamp system is disconnected from the testing machine.

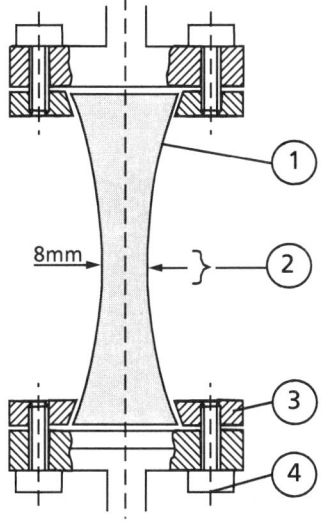

Fig. 10.15. Rod-like specimen for tension-compression tests.
(1) specimen
(2) gauge length
(3) split cone
(4) screw

10.8.2 Measurement of Crack Propagation

For the mechanical testing machine, described in the preceeding section, a high load rate is useful. For investigation of crack propagation, by static load, however, very small load rates are required. For most strain-controlled crack propagation measurements, in addition, the stiffness of the apparatus must be much higher than that of the polymeric specimen. The reason is, that no strain from the apparatus should be superimposed to the sample strain or crack opening displacement during unstable crack propagation. When measuring fracture energy (critical energy release rate) the strain position of the apparatus has to be stopped immediately when unstable crack propagation starts. Otherwise energy fed in from the apparatus falsifies the measurement. An apparatus obeying these requirements is shown in Figure 10.16. The specimen is loaded by a lever arm, which is fixed in a very stable frame. There are combined several advantages:

- accurately controllable strain
- high stiffness of the specimen fixation by a stable frame,
- a thin pull rod with low heat conduction transmits the load which is enhanced by the lever arm.

Cooling by a bath cryostat with a liquid or gaseous environment is sufficient for most cases. The crack opening displacement is detected by an extensometer. If direct observation of crack propagation is required a window through the cryostat wall and an optical system can be installed.

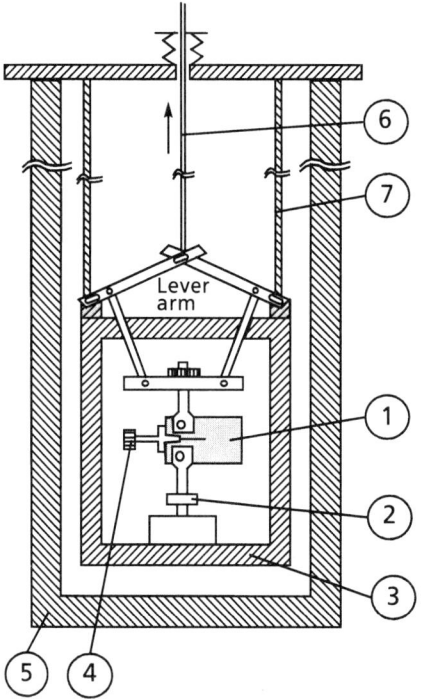

Fig. 10.16. Apparatus for measuring crack propagation at low temperatures.
(1) compact tension specimen with an initial crack
(2) piezoelectric quartz
(3) stable frame
(4) extensometer
(5) bath cryostat
(6) pull rod
(7) holder.

For investigation of crack propagation, in most cases, compact tension (CT) samples are used. Since polymers are brittle at low temperatures, special type of triangular shaped specimens are applied. These so-called Chevron specimens have the advantage, that even an uncontrolled crack propagation can be arrested within the specimen in many cases. A top view of a Chevron specimen cut in the middle plane is shown in Figure 10.17.

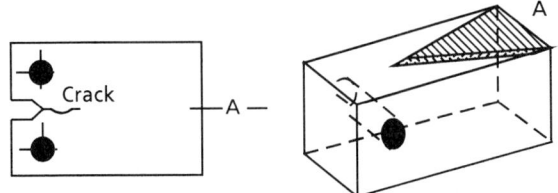

Fig. 10.17. Side view and middle plane view of a Chevron specimen.

10.9 SUMMARY

The aim of this chapter was the description of several test apparatus modified for cryogenic use. These principles can be transposed to other devices. Main steps for designing low-temperature devices are:

- choice of the cooling system,
- fixation of the specimen within the cryostat,
- data detection at low temperatures,
- reduction of heat leaks (e.g., heat input to sensors or specimens),
- improvement of the accuracy of measurement if a property gets very small at low temperatures,
- choice of an adapted shape of specimens for use at cryogenic temperatures.

In many cases a compromise or an optimization has to be made in order to satisfy different requirements.

10.10 GENERAL REFERENCES

(1) Scott, R. B.: Cryogenic Engineering Reprinted 1988 by Met.-Chem. Research Inc., Boulder, Colorado
(2) Hudson, R. P.: Principles and application of magnetic cooling, North-Holland Publishing Company, Amsterdam - London (1972)
(3) Pobell, F., Matter and Methodes at Low Temperatures; Springer-Press, 1991.
(4) Experimental Techniques in Condensed Matter Physics; Eds. R. C. Richardson and E. N. Smith, Addison Wesley, Redwood City, Col. (1988)
(5) White,G,K; Experimental Techniques in Low-Temperature Physics; Oxford Science Publications (1979)

POLYMERS AS MATRIX FOR COMPOSITES

Contents

11.	Polymers as matrix for composites	243
11.1	Mixing rules	243
11.2	Lamination theory	245
11.3	Fiber-matrix bond	246
11.4	Thermal stresses and relaxation	246
11.5	Fracture properties	247
11.6	Matrix materials for fiber composites used at cryogenic temperatures	248
11.7	Summary	250
11.8	References	250

11. POLYMERS AS MATRIX FOR COMPOSITES

An important application of polymers is their use in composites, the filler being fibers or particles. Especially fibers are the reinforcing elements of a tough but weak polymer matrix.[1] It is the goal of several rules or theories to predict the composite properties from those of the components. The important question arises whether or not bulk properties of polymers can be used in composite analysis. Several matrix properties are changed by the constraints of the surrounding fillers. Mixing rules and lamination theory are employed rather successfully, especially for calculating elastic and thermal properties. The calculation of several composite properties will be considered briefly in the following sections.

11.1 MIXING RULES

They are used to predict properties of composites which consist of two (or more) components of known concentrations and properties. The application of these rules is restricted to isotropic properties without tensorial features. For example, the thermal conductivity in most cases cannot be predicted by a linear mixing rule. Also, if there are interactions of filler and matrix, a linear mixing rule must at least be modified. For example, internal stresses between filler and matrix, induced by temperature variations, may influence the properties, such as thermal expansion. Boundary scattering of phonons at particle/matrix interfaces influences the thermal conductivity at very low temperatures (see Subsection 5.4.1). But there are several properties which can be predicted from mixing rules. The following notations are used: subscripts c for composites, f for fiber and m for matrix; ϕ is the filler content by volume.

11.1.1 Specific Heat

For any composite the specific heat can be calculated from the specific heat c_m of the matrix and c_f of the filler (fiber or particulate filler).

$$c_c = \phi\, c_f + (1 - \phi)\, c_m \tag{11.1}$$

The bulk property c_m of the matrix is applicable and does not change within a composite.

---

[1] This is different from reinforcing a strong but brittle ceramic matrix. Toughness of fiber/ceramic composites is increased by friction between matrix and fibers.

11.1.2 Permittivity

The permittivity ε_c of a composite with particulate filler can be calculated only roughly by a mixing rule since the electrical field is not uniform.

$$\varepsilon_c \approx \phi\, \varepsilon_f + (1-\phi)\, \varepsilon_m \tag{11.2}$$

For fiber composites, however, the permittivity is anisotropic, and only the tensor component perpendicular to the fiber layers obeys roughly a mixing rule. The matrix property within a composite does not change.

11.1.3 Mechanical- and Dielectric Damping

For isotropic, particle filled materials and to some extend for unidirectional fiber composites a mixing rule is applicable

$$\tan\delta_c = \phi\, \tan\delta_f + (1-\phi)\, \tan\delta_m \tag{11.3}$$

The loss-factor $\tan\delta_f$ of most reinforcing fillers is negligible and the matrix determines the damping behavior. Only Kevlar fibers contribute appreciably to the mechanical damping. The damping behavior of most other fiber composites is determined by the relative amount of the matrix. The loss-factor of uni-directional fiber composites is somewhat dependent on the direction of measurement (parallel or perpendicular to the fiber direction), but is determined mainly by the matrix damping. Roughly a mixing rule with bulk properties can be applied. It should be mentioned that cracks in composites increase the mechanical damping behavior due to friction. (The increase of crack density can be determined by damping measurements, e.g., at processes where crack propagation occurs).

11.1.4 Moduli

The moduli of particulate composites can be calculated only roughly by a mixing rule. The reason is the complex stress distribution under uniaxial load. (Under hydrostatic loads a mixing rule is valid). The same is true for the modulus of fiber composites perpendicular to the fiber direction.

11.1.5 Thermal Expansion

Even for powder filled polymers a linear mixing rule yields only a rough approximation since thermal stresses are induced by temperature variations. Because of the difference in thermal expansion of filler and matrix the thermal stresses deform the matrix, and thus influence the thermal expansion of a composite. The thermal stresses depend on the filler shape. Spherical and cubic particles, for example, exhibit a different expansion behavior. These dependences are shown in Figure 11.1. The curves (b) and (c) are calculated by stress analysis [11.1]. Experimental results are only roughly in agreement with a linear mixing rule (curve a). For fiber composites the anisotropic expansion behavior is a topic of the lamination theory; only the expansion perpendicular to the fiber plane can be estimated by a mixing rule.

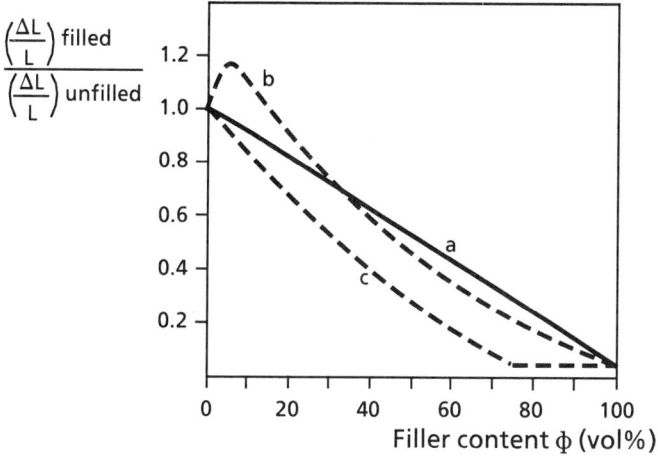

Fig. 11.1. Coefficient of thermal expansion of a particle filled epoxy resin.
a) linear mixing rule;
b) cubic fillers;
c) spherical fillers.

11.2 LAMINATION THEORY

This theory is very successful in predicting anisotropic thermal and mechanical deformation properties of fiber composites [11.2]. The first step concerns the calculation of unidirectional fiber composites from the properties of fiber and matrix. The stress analysis works mainly on two principles:

- Compatibility of all vector components { } of composite strains with those of the fibers and matrix in a composite

$$\{\varepsilon\}_c \equiv \{\varepsilon\}_f \equiv \{\varepsilon\}_m \tag{11.4}$$

- Compensation of internal stresses

$$\sum \sigma_{ij} = 0 \tag{11.5}$$

The second step takes into account thermal expansion at temperature variations. This is done by the components of thermal expansion a_m and $\{a\}_f$. The strain components of the matrix and fiber are given by

$$\{\varepsilon\}_m = [c]_m \{\sigma\}_m + a_m \Delta t \tag{11.6}$$

$$\{\varepsilon\}_f = [c]_f \{\sigma\}_f + \{a\}_f \Delta t \tag{11.7}$$

$[c]_m$ and $[c]_f$ are the compliance tensors of the matrix and fiber; $\{\sigma\}_m$ and $\{\sigma\}_f$ are the internal stresses. Using Eqs.11.4 to 11.7 it is possible to calculate the moduli and thermal expansion coefficients of unidirectional fiber composites.

The properties parallel and perpendicular to the fiber direction are called orthotropic properties. They can be used to calculate the properties of angle plies, when unidirectional fiber layers are stacked sandwich-like in $\pm\theta$ directions. The orthotropic properties are transformed by a matrix T into the properties in a coordinate system, which is rotated by an angle θ (see Subsection 4.2.4).

$$\{\}_\theta = T\{\}_{ortho} \tag{11.8}$$

The parenthesis { } can be filled with components of any property, e.g., modulus, thermal expansion coefficient.

$$T = \begin{bmatrix} \cos^2\theta & \sin^2\theta & \sin 2\theta \\ \sin^2\theta & \cos^2\theta & -\sin 2\theta \\ \frac{-1}{2}\sin^2\theta & \frac{1}{2}\sin 2\theta & \cos^2\theta - \sin^2\theta \end{bmatrix} \quad (11.9)$$

The compatibility is now applied to $+\theta$ and $-\theta$ laminates

$$\{\varepsilon\}_c \equiv \{\varepsilon\}_{+\theta} \equiv \{\varepsilon\}_{-\theta} \quad (11.10)$$

The internal forces between $\pm\theta$ laminates cancel each other, if laminates are glued together.

$$\sum \{\sigma_{ij\Theta}\} = 0 \quad (11.11)$$

More generally, any angle ply with $\pm\theta_1$, $\pm\theta_2$ .... can be treated in this way. By means of the lamination theory moduli, mechanical deformations, thermal expansions and permittivities can be calculated. In an approximation the bulk properties of the matrix are used to calculate the composite properties [11.3; 11.4].

11.3 FIBER-MATRIX BOND

The lamination theory assumes excellent bond between fiber and matrix or between layers (laminates). The bond strength mainly depends on the manufacturing process and on the fiber-matrix combination. Fibers usually are coated with a special finish which is adapted to the desired matrix material. PEEK is a matrix polymer with a high bond strength because it crystallizes on the fiber surface. When a composite is cooled to low temperatures the original bond strength is less important since it is increased by shrinkage of the matrix onto the fiber surface. The effect of shrinkage, however, is smallest for composites with carbon- and Kevlar fibers, which show a relatively high transverse thermal expansion[11.3].

A measure of the bond strength is the interlaminar shear strength. Its maximum value (ideal bond) is given by the shear strength of the matrix, which is about 200MPa at 4.2K. Usually, short beam bending tests are used for a rough determination of the interlaminar shear strength. The load deflection curves at this test exhibit at low temperatures a saw tooth shape which reflects successive delaminations. At elevated temperatures and for thermoplastic matrices, however, the diagrams are smeared out.

11.4 THERMAL STRESSES AND RELAXATION

Temperature change induces thermal stresses due to the different thermal expansion of filler and matrix. Each isotropic polymer matrix exhibits a much higher thermal expansion than most fillers or fibers (at least in fiber direction).

The thermal prestress on the matrix is crucial, especially at low temperatures where polymers become brittle. This reduces the effective strain to failure and is a source of microcracks in the matrix. Actually, thermal stresses are accumulated starting from the elevated manufacturing temperatures. From the relaxation behavior of the bulk material it was expected that the thermal prestress would vanish after a sufficient period of time. This proved to be true in part only. Multiaxial stresses and steric hindrance in a composite reduce the relaxation strength. Even for thermoplastic matrices with a rather high relaxation strength, the thermal prestress is reduced by a small portion only even at RT. This can be demonstrated on nonsymmetrically stacked crossplies. When cooling down from the manufacturing temperature (650K for PEEK) the composite plane becomes warped. The curvature does not relax appreciably even if it is helt flat for several days at RT [11.5]. At low temperatures the situation is even worse.

11.5 FRACTURE PROPERTIES

It is difficult to predict the fracture properties of composites since various fracture processes are involved which depend on the fiber- matrix system. At low temperatures the thermal stresses complicate the prediction. Linear mixing rules or the lamination theory are not sufficient. Special theories have been developed to predict the static fracture strength. The most critical strength is always the tensile strength [11.6].

Static Tensile Strength of Polymer-Particle Composites. As already mentioned, uniaxial load causes complicated stress distributions and stress concentrations within the particle-matrix system. The fracture strength found for polymers as bulk material is thus only an upper limit of the composite tensile strength. Even more reduced is the fracture strength, if a composite is cooled, and thermal stresses are superimposed. Since thermal expansion of polymers is much larger than that of most rigid particle fillers, the matrix is prestressed around the surface of each particle when cooled. There exists for each temperature decrease a critical particle diameter, above which the tensile strength is reduced. The critical particle diameter can be calculated from the thermal and mechanical properties of the filler and the matrix (measured as bulk properties). For composites with very small particles the thermal stresses are distributed more uniformly and do not reduce fracture strength [11.7]. This is not true for composites with large particles for the same filling fraction.

Static Tensile Strength of Fiber Composites. Only matrix dominated properties (e.g., shear strength or strength perpendicular to the fiber direction) are influenced by the type of matrix polymers applied.

Fatigue Behavior. The fiber dominated components of static strength are almost not influenced by the matrix type. This is not true for the fatigue behavior, not even for unidirectional fiber composites in fiber direction. This is demonstrated in Figure 11.2 for tensile threshold fatigue loading on unidirectional fiber composites which consist of the same fibers but have different matrices. Composites with brittle epoxy matrices show a much higher fatigue endurance limit than those with tough thermoplastic ones (e.g., PEEK; PC).

During fatigue cycling a much higher density of microcracks is accumulated in the rigid matrix than in the tough one. Many cracks, however, are less dangerous since their stress concentrations are smaller; the stress concentration under tensile load is distributed over many cracks [11.8]. The fatigue behavior of a matrix within a fiber composite is much different from the bulk property.

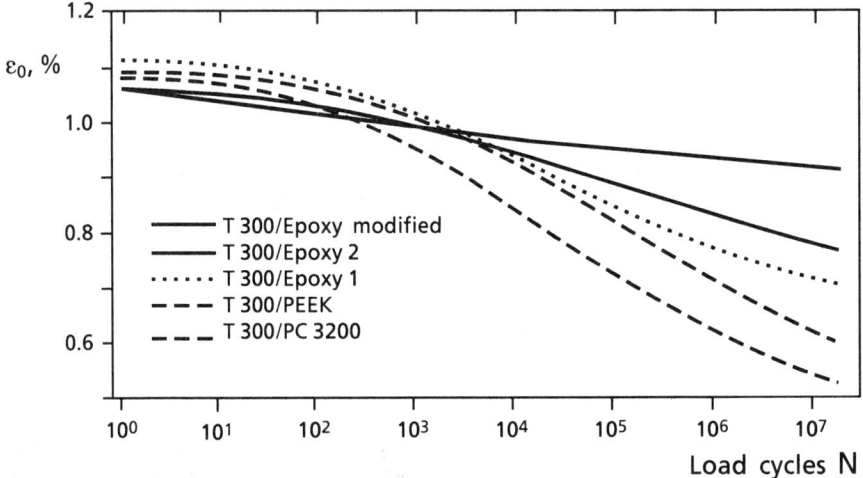

Fig. 11.2. Fatigue curves of unidirectional carbon fiber composites with different matrices. The strain live at tensile threshold load cycling at 77K is plotted versus load cycles.

11.6 MATRIX MATERIALS FOR FIBER COMPOSITES USED AT CRYOGENIC TEMPERATURES

It is difficult to give a general rule allowing desired composite properties to be achieved. Costs of materials and economic processing methods restrict the choice of matrices furthermore. The following parameters are important for the proper selection of a matrix:

- Desired composite properties.
- Processing method.
- Costs of the matrix material.
- Recycling

These parameters will be briefly treated below.

Desired Composite Properties. The mechanical composite properties at low temperatures are influenced by the matrix. Epoxy resins are more brittle than thermoplastic polymers. This is an aspect for using thermoplastic polymers as matrix. PEEK, PSU, PES, PEI, PP are some of the commonly used thermoplastic matrices. The fracture strain of matrices at 4.2K and of fibers are compiled in Table 11.1. It can be seen that fracture strain of thermoplastic matrices is larger than that of carbon-, Kevlar- or ceramic fibers, even if the thermal prestrain on the matrix ($\approx 1\%$ at 4.2K) by cooling is subtracted.

Table 11.1. Fracture strains at 4.2K.

Epoxy resins	≈ 2 %
Thermoplastic matrices	3 – 4 %
Fiber glass	3 %
Carbon fibers	1.2 – 2.2 %
Kevlar fibers	1.5 – 2 %
Ceramic fiber	< 1.4 %

A thermoplastic polymer matrix in a carbon fiber composite does not break prior to the composite. This is not quite true for fiber glass composites and not true when epoxy resin matrices are applied. The thermoplastic matrix PEEK, for example, exhibits a high bond strength and a rather high fracture strain at low temperatures. At static tensile loading of unidirectional (UD) composites with a PEEK matrix nearly no microcracks have been observed. Matrix cracking generally is more serious for crossplied composites.

At cyclic loading the situation is different especially for carbon fibre composites. Carbon fibres themselves exhibit a very high fatigue endurance limit. They are, however, sensitive to the behavior of the matrix. As already shown in Figure 11.2 the fatigue endurance limit (strain) of UD composites is higher with epoxy matrices, even when they exhibit a higher crack density. Best results, not only at low temperatures, are achieved with a modified epoxy matrix, whose chains are toughened by inserting thermoplastic components between cross-link points [11.8]. UD carbon composites with a PEEK matrix, by contrast, show a low fatigue endurance limit. Thus, the proper matrix selection depends on the type of loading. Crossplies (or angleplies) are less sensitive to the fatigue behavior of the matrix.

There are various environmental influences, which may degrade the matrix properties. Radioactive irradiation or chemicals are some examples. The matrix polymers with the highest resistivities are polyimides (Vespel, Kerimid), followed by PEEK, PEI and special epoxies (aromatic amine resins) [11.9]. The costs of polyimides are very high. PEI and epoxies are less expensive.

Processing Methods There are several types of polymers which are normally used as matrix materials.

- Epoxy resins: In the liquid state they can be used as impregnation material for wet lay up technique. As prepreg material epoxy resins are less favorable because the precured epoxy coating of the fiber fabrics can be stored only over a short time (some weeks).

- Thermoplastic matrices: They are less favorable for a wet lay up technique but are suited as prepreg material. The prepreged laminates can be stored for any time and processed by hot pressing.

11.7 SUMMARY

In several cases it is possible to calculate the composite properties from those of the components. Polymers in composites, however, may exhibit a behavior different from that of the bulk material. Table 11.2 summarizes whether or not polymer bulk properties can be applied in composite analysis; some restrictions for the applicability are included.

Table 11.2: Applicability of polymer bulk properties

Properties	Bulk properties	Restrictions
Specific heat	yes	
Thermal conductivity	yes	boundary effects below 5K.
Thermal expansion	yes	
Permittivity	yes	
Mechanical and dielectric damping	yes	microcracks may increase the values.
Relaxation	no	multiaxial stresses and steric hindrances by the fibers.
Moduli	yes	
Fracture properties	no	multiaxial and internal stresses, stress concentrations.

The choice of a matrix depends on the desired composite properties, the processing method and the environmental influences. Costs and recycling problems may cause further restrictions on the choice of a proper matrix.

11.8 REFERENCES

11.1 Hartwig, G.; A. Puck and W. Weiss; Mat. Scr. Eng. ; Vol 22; p. 270;
11.2 Shapery, R.A.; J. Comp. Mat., Vol. 2; p. 380; (1968)
11.3 Hartwig, G.; Cryogenics 28, (1988), p. 255
11.4 Hartwig, G.; Cryogenics 28, (1988), p. 216
11.5 Ahlborn, K.; KfK-Report 4487, (1989), p. 30
11.6 Composites, Vol. 1; p. 236; Engineered Materials Handbook; ASM International; Metal Park, Ohio; (1987)
11.7 Lange, F.F.; in Composite Materials, Vol 5; p. 26; Eds. L. J. Broutman and R. H. Krock; Academic Press;New York; (1974)
11.8 Pannkoke, C.; KfK-Report 5112, (1992) and Cryogenics, Vol. 31, (1991), p. 248
11.9 Hartwig, G.;KfK-Report 03.01. P4.2A, (1985) Karlsruhe, Germany

PREFACE TO THE APPENDIX

Data Base

It is a difficult task to compile data on materials, which differ slightly from one another. A chemical formula characterizes only the basic structure of a polymer. Processing methods, additives and post treatments influence the properties to some extent. These influences, however, are in most cases less pronounced at low temperatures. The low-temperature data are less sensitive to the details in composition or morphology. The data at room temperature should be considered as approximate values only. In this data base no trade names of polymers are given since most of them are used in the respective national nomenclature. The data without references have been measured in the author's laboratories, some data have not yet been published elsewhere. The collaboration of all coworkers is greatly acknowledged.

Appendix 1A

Polymer	Abbreviation	Composition of a Repetitive Unit
Polyethylene (High density) (Low density)	PE (HDPE) (LDPE)	$\left[\begin{array}{cc} H & H \\ -C-C- \\ H & H \end{array}\right]$
Polyvinyl chloride	PVC	$\left[\begin{array}{cc} H & H \\ -C-C- \\ H & Cl \end{array}\right]$
Polycarbonate	PC	$\left[-O-\bigcirc-\underset{CH_3}{\overset{CH_3}{C}}-\bigcirc-O-\underset{\parallel}{\overset{O}{C}}-\right]$
Polystyrene	PS	$\left[\begin{array}{cc} H & C_6H_5 \\ -C-C- \\ H & H \end{array}\right]$
Polytetrafluoro-ethylene	PTFE	$\left[\begin{array}{cc} F & F \\ -C-C- \\ F & F \end{array}\right]$
Polyimide	PI	(imide structure with two fused rings linked via N to diphenyl ether)
Polyetherimide	PEI	(etherimide structure)
Polypropylene	PP	$\left[\begin{array}{cc} H & H \\ -C-C- \\ H & CH_3 \end{array}\right]$
Polysulfone	PSU	$\left[-O-\bigcirc-\underset{CH_3}{\overset{CH_3}{C}}-\bigcirc-O-\bigcirc-\underset{\parallel O}{\overset{O}{S}}-\bigcirc-\right]$

Appendix 2A

Polymer	Abbreviation	Composition of a Repetitive Unit
Polyoxymethylene	POM	$\left[\begin{array}{cc} H & H \\ -C-O-C-O- \\ H & H \end{array}\right]$
Polymethylmethacrylate	PMMA	$\left[\begin{array}{cc} H & CH_3 \\ -C-C- \\ H & C=O \\ & O \\ & CH_3 \end{array}\right]$
Polyether sulfone	PESU	$\left[-\bigcirc-\underset{\underset{O}{\overset{O}{\parallel}}}{S}-\bigcirc-O-\right]$
Polyetherether ketone	PEEK	$\left[-\bigcirc-\underset{\overset{\parallel}{O}}{C}-\bigcirc-O-\bigcirc-O-\right]$
Polyether ketone	PEK	$\left[-\bigcirc-\underset{\overset{\parallel}{O}}{C}-\bigcirc-O-\right]$
Liquid Crystalline Polymer	LCP (Vectra A 950)	$\left[-O-\bigcirc-\underset{\overset{\parallel}{O}}{C}-\right]_{73\%} + \left[-O-\bigcirc\bigcirc-\underset{\overset{\parallel}{O}}{C}-\right]_{27\%}$
Polyamide 6	PA 6	$\left[\begin{array}{cccccc} H & H & H & H & H & H \\ N-C-C-C-C-C-C- \\ H & H & H & H & H & O \end{array}\right]$
Polyamideimide	PAI	Structure with imide and amide linkages; R: small molecular units.
Epoxy (semi-flexible)	CY 221 / HY 979	Structure with □: several units from PC

Appendix 3 A
Table 3 A: Specific heat $c_p \cdot 10^{-3}$, J/gK

T,K	PS	PVC	PMMA	PETP	PTFE$_{sc}$	PC	EP	PP$_a$	PP$_c$	HDPE$_a$	HDPE$_c$	POM$_{sc}$
10	31	29	17	64	26	24	22	25	17	25	8	13
20	102	95	80	119	79	102	85	120	87	98	52	80
30	166	160	147	180	126	159	170	205	165	191	134	175
40	219	214	214	251	170	210	220	281	236	276	226	258
50	266	260	280	350	210	255	270	354	304	346	329	325
77	370	360	420		310	348	400	543	472	550	540	450
100	453	432	550	438	392	439	480	684	613	682	682	550
150	617	562		620	550	636		944	990	936	860	710
200	798	688	920	800	677	832	1000	1227	1110	1190	1030	835
250	988	816		983	780	1028		1416	1314	2070	1220	985
300	1210	944		1170	870	1228	1300*	2077	16520	2270	1440	1170
400	1909											
Ref.	1)	2)	3)	4)	5)	6)	7)	8)	8)	9)	9)	9)

For semicristalline polymers the extrapolated values of the amorphous (a) and crystalline (b) phases are given. The values for any crystalline content can be calculated by a mixing rule.

sc) semicrystalline, (ca. 60%–70% crystalline content).

*) Epoxy: Cy 221, Hy 979 (Ciba-Geigy)

References

1. Gaur, U., B. Wunderlich; J. Phy. Chem. Ref. Data, Vol. 11; p.322; (1982) and Kirkpatrick, D.E., L. Jodoreits, B. Wunderlich; J. Polym. Sci.; Phys. Eds, Vol.24; p.45; (1986)
2. Gaur, U., S.F. Lau, B. Wunderlich; J. Phy. Chem. Ref. Data, Vol.12; p.52; (1983)
3. Reese, W.; J. Macromol. Sci. Chem. A 3 (7); p.125; (1969)
4. Gaur, U., B. Wunderlich, J. Phys. Chem. Ref. Data, Vol. 12; p.75; (1983)
5. Loufakis, K., B. Wunderlich; Polymer Vol. 26 ; p.1875; (1985) and Boyer, J.D. et al.; J. Non-Cryst. Solids 55; p.913; (1983)
6. Gaur, U., S.F. Lau, B. Wunderlich; J. Phys. Chem. Ref. Data, Vol. 12; p.95; (1983)
7. Hartwig, G.; Prog. in Colloid and Polym. Sci. 64; p.56; (1978)
8. Grebowicz,J., S.F. Lau, B. Wunderlich; J. Polym. Sci. Polym. Symp. 71; p.19; (1984)
9. Engeln J, Thesis (1983) TU-Berlin and Ref.[3.3].

Appendix 4 A

Table 4 A: Thermal expansion.

Polymer	$-\Delta L/L_0, \%$ 293-77 K	$-\Delta L/L_0, \%$ 293-4,2 K	$\alpha \cdot 10^{-6}, K^{-1}$ 30 K	$\alpha \cdot 10^{-6}, K^{-1}$ 100 K	$\alpha \cdot 10^{-6}, K^{-1}$ 200 K	$\alpha \cdot 10^{-6}, K^{-1}$ 293 K	Ref.
PEEK	0.9	1.0	10	34	43	47	8)
PESU	0.9	1.1	13	32	46	54	8)
PETFE	1.4	1.6	16	43	65	87	8)
PETP	1.1	1.3	13	40	52	61	8)
PC	1.2	1.4	16	42	56	65	8)
	-	-	18	39	56	67	7)
PSU	1.0	1.1	13	37	48	55	8)
PMMA	0.9	1.0	10	31	40	62	8)
PS	1.3	1.5	18	52	62	71	8)
PVC (grey)	1.0	1.1	19	28	54	74	8)
(red)	1.1	1.3	16	38	53	79	8)
PVDF	1.4	1.5	15	39	56	118	8)
PI (Kerimid)	0.8	0.9	12	27	39	46	8)
PI (Vespel)	0.8	0.9	8	27	36	40	8)
PI	-	-	8	14	18	22	7)
PEI	0.8	1.0	15	29	40	53	8)
PAI	0.7	0.8	10	24	33	37	8)
PA 12	1.5	1.6	8	43	81	110	8)

Appendix 4 A

Table 4 A: Thermal expansion (continued).

Polymer	$-\Delta L/L_0$, % 293-77 K	$-\Delta L/L_0$, % 293-4,2 K	$\alpha \cdot 10^{-6}$, K^{-1} 30 K	$\alpha \cdot 10^{-6}$, K^{-1} 100 K	$\alpha \cdot 10^{-6}$, K^{-1} 200 K	$\alpha \cdot 10^{-6}$, K^{-1} 293 K	Ref.
Epoxy (Cy221/Hy979)	1.0	1.1	12	32	47	84	8)
Epoxy (DGEBA diam.)	0.9	1.1	-	-	-	-	1)
Epoxy (rigid)	1.0	1.1	-	-	-	-	2)
(semiflexible)	1.3	1.5	-	-	-	-	
HDPE	2.0	2.1	10	49	95	158	8)
PE	-	-	14	51	90	128	3)
PE	-	-	22	55	88	150	4)
PE	-	-	-	53	88	142	5)
PE	-	-	16	50	89	130	6)
PTFE	1.7	1.8	12	50	83	112	8)
POM	1.4	1.5	14	42	67	98	8)
PP	1.2	1.3	17	52	55	110	8)

Appendix 4A: Thermal Expansion

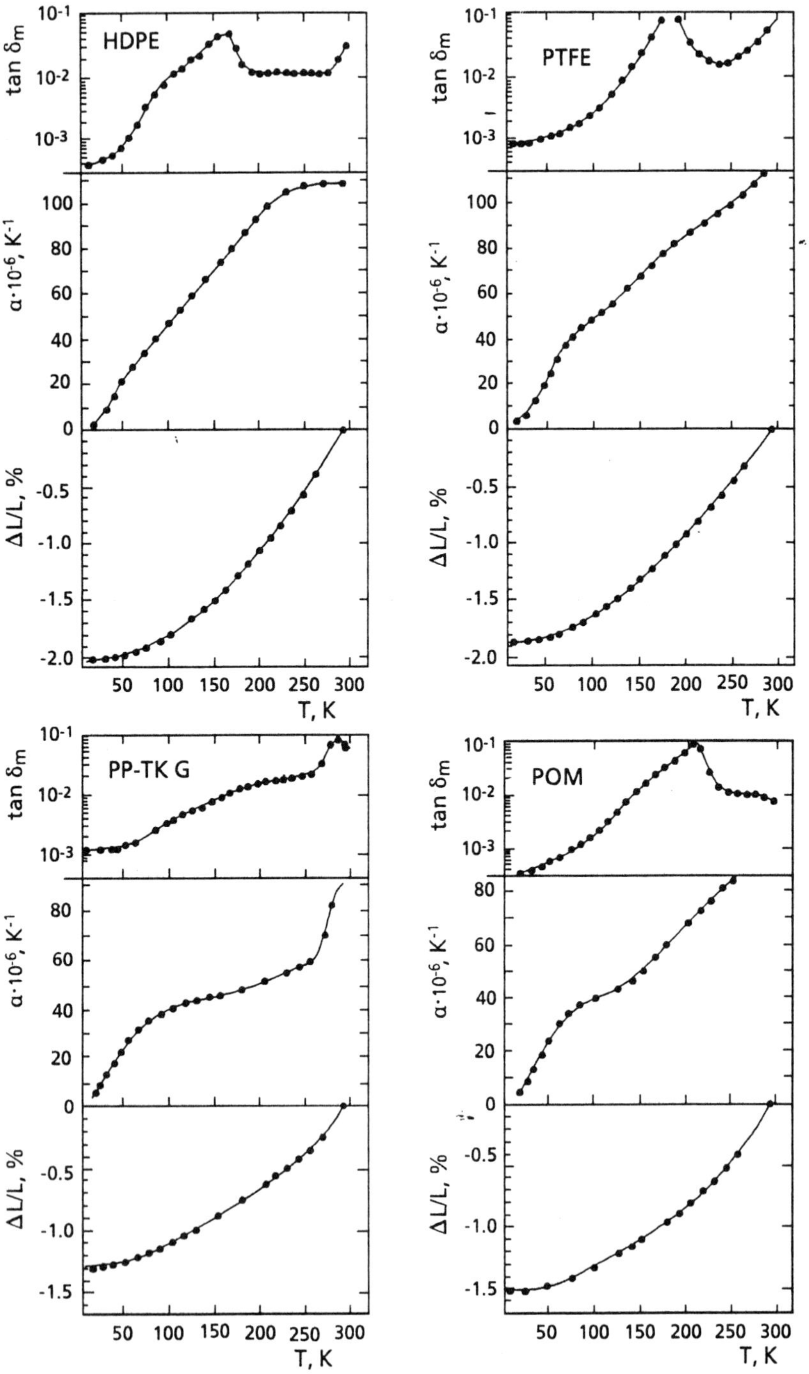

Thermal expansion coefficient α; integral thermal expansion ΔL/L; loss-factor tanδ.

Appendix 4A: Thermal Expansion

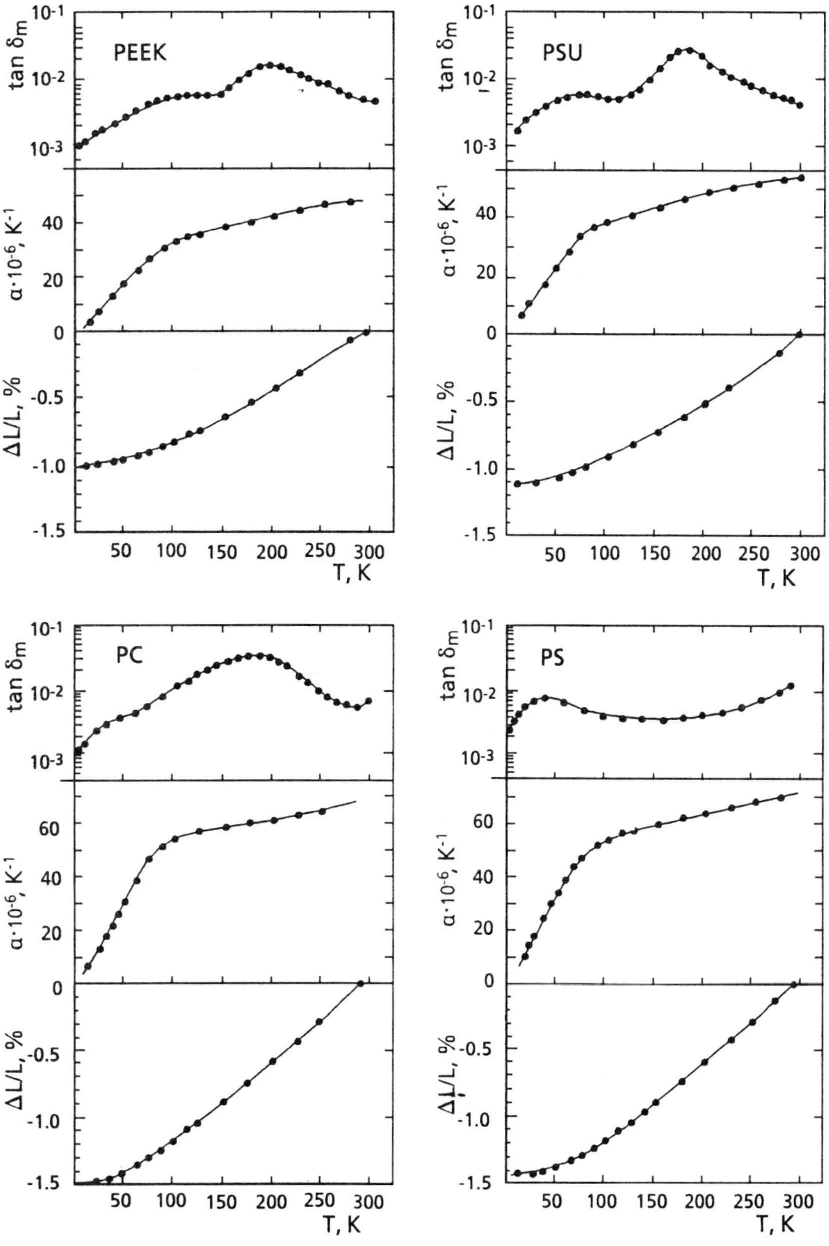

Thermal expansion coefficient α; integral thermal expansion ΔL/L; loss-factor tanδ.

Appendix 4A: Thermal Expansion

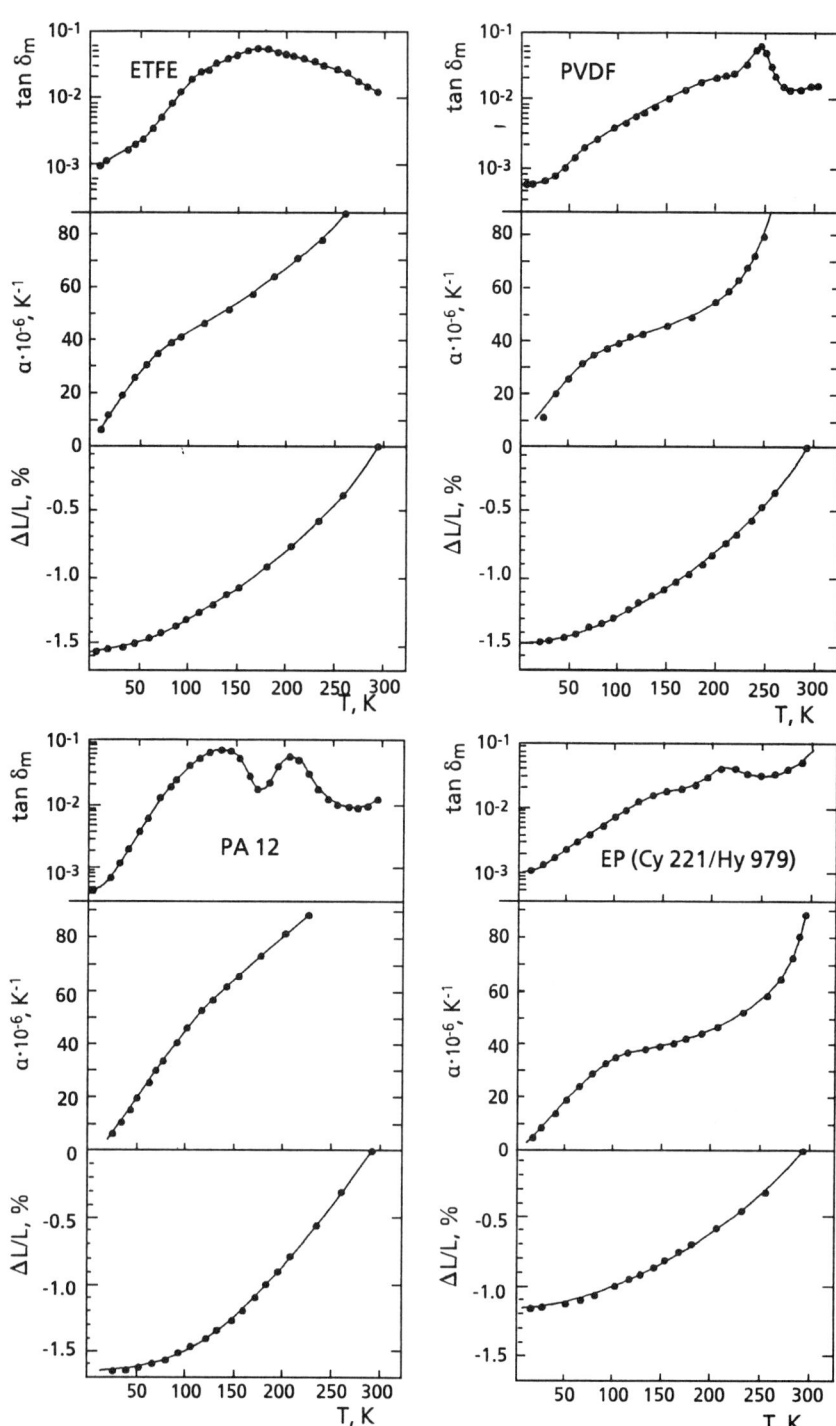

Thermal expansion coefficient α; integral thermal expansion ΔL/L; loss-factor tanδ.

APPENDIX

References for 4A

1) Evans, D.; Morgan, J. T.; Nonmetallic Materials and Composites at Low Temperatures 2; pp 73-87; (1982)

2) Evans, D.; Morgan, J. T.; Proceedings of the ICMC, Kobe, Japan, pp 446-450; (1982)

3) White, G.K.; Choy, C. L.; Journal of Polymer Science, Vol. 22; pp 835-846; (1984), (PE crystallinity 0,8)

4) Dadobajev, G.; Slusker, A.I.; Sov. Phys. Solid State 23, 7, pp 1131-1135; (1981)

5) Perepechko, I.; Low Temperature Properties of Polymers; p. 86; (1977); Pergamon Press

6) Engeln, I.; Meissner M.; Nonmetallic Materials and Composites at Low Temperatures 2, pp 1-25; (1982) (PE:crystallinity 0.77)

7) Roe, J. M.; Simha, R.; Int. Journal of Polymeric Material Vol. 3; pp193-22; (1974)

8) Measured at the author's laboratory.

Appendix 5A

Table 5A: Thermal conductivity; $\kappa \cdot 10^{-2}$; W/(m,K).

Polymer	1K	2K	4.2K	10K	70K	100K	150K	300K	Ref.
HDPE	0.3		2.9	9.0	40	45		40	1); 2);
MDPE		0.7		6.0					3)
PTFE	0.3			9.5				25	3)
PP		0.5	1.2	3.4	16	18	20	22	11)
POM		1.2	3.0	9.3	49	50	49	40	11)
PETP (amorphous)		3.0	3.5	4.5	15				4); 10)
(cryst. 25 %)		0.7	1.3	3.0	17				
PC	2.0	2.5	3.1	4.0	13	15			11)
PVC		2.0	2.7	4.0	12	13	13	14	11)
PS	1.1		2.9	3.0		5			13)
PMMA	2.0			6.0		16		20	8)
	3.0			7.5		20			7)

Appendix 5A

Table 5A: Thermal conductivity ; (continued)

Polymer	1K	2K	4.2K	10K	70K	100K	150K	300K	Ref.
PI									
(Kerimid)		2.5	3.8	4.0		9	12	23	9)
(Vespel)			1.1						
Epoxies			4.8¹⁾	6.0		13		22	
			7.1²⁾	8.5		16		25	
ISR F4				10.0	19				1)
Scotchcast			5.4	5.9		13		21	12)
(SL8)	3–4		5.0	10.0					5)
			7–9						5)

1) cross - link distance ≈ 1,5 nm
2) cross - link distance ≈ 12 nm

References for 5A

1) Van de Voorde; CERN 77-8 ISR-BOM 8 Jan.; (1977).

2) Engeln, J. and M. Meissner; Nonmetallic Materials and Composites at Low Temperatures; Plenum Press; New York (1982); p.4.

3) Perepechko, I.I.; Low-Temperature Properties of Polymers, Pergamon Press; (1980), p. 57.

4) Greig, D. and M. Sahota, ; J. Phys. C: Solid State Phys. 16; p.1051; (1983).

5) Rosenberg, H.M.; Nonmetallic Materials and Composites at Low Temperatures 2: p. 181; (1980); Plenum Press.

6) Stephens, R.B.; Phys. Rev. B, Vol. 8, Nr. 6, p. 2896; (1973).

7) Finnlayson, D.M. and P. Mason; J. Phys. C: Solid State Phys., (1984).

8) Finnlayson, D.M., P. Mason; J.N.Rogers, and D. Greig; J. Phys. C, solid State Phys. 13; p. 185, (1980)

9) Claudet, G.; F.Disdier and M. Locatelli; Nonmetallic Materials and Composites at Low Temperatures; p. 131, Plenum Press (1978).

10) Choy, C.L.; G.L Salinger and Y.C. Chiang; NBS Spec. Publ. 301; p. 567 (1969).

11) Choy, C.L.; and D. Greig; J. Phys C. Solid State Phys. Vol. 10; (1977); p. 169.

12) Evans, D. and J.T. Morgan; Proceedings ICMC, Kobe, Japan; p. 286, Butterworths Press (1982).

13) Hager, N.E. jr.; Rev. Sci. Instr. 31; p. 177; (1960).

Appendix 6A
Table 6A: Loss-factor tanδ at 6-10Hz

Polymer	tanδ×10⁻³			
	4.2K	77K	150K	293K
amorphous				
PEA[1]	7.0	14.0	20	>200
PMA[1]	5.0	6.0	20	>200
PS	2.0	4.5	4	10
PSU	2.0	6.0	20	4
PEC	2.0	8.0	50	3
PEMA	2.0	9.0	10	120
PBMA	0.2	4.4	2	2
PC	1.2	6.0	26	8
PVC	1.0	6.0	12	13
PEEK	1.0	4.0	7	5
PMMA	0.6	2.0	7	90
PEI	1.6	2.4	4	12
PAI	1.3	3.0	7	12
PI (Vespel)	1.1	5.2	7	11
cross-linked				
PI (Kerimid)	1.4	3.5	9	15
EP[2]	1.0	4.0	16 to 30	10 to 65
semicrystalline				
ETFE	1.0	7.0	50	12
PTFE	0.8	1.7	35	80
FEP	0.6	3.0	40	12
PVDF	0.6	3.3	15	22
PA 11	0.5	2.0	50	11
PBTP	0.3	7.0	35	4
PA 6.10	0.5	13.0	33	7
LDPE	0.5	10.0	52	65
PFA	0.5	2.7	50	15
HDPE	0.3	5.0	54	26
POM	0.3	1.0	16	8
PP	1.2	2.0	10	60

Nomenclature of special polymers or copolymers:

PEA	: Polyethylacrylate	ETFE	: copolymer PE / PTFE
PMA	: Polymethylacrylate	FEP	: copolymer PTFE / hexafluorpropylene
PEC	: postchlorided PE	PVDF	: Polyvinyl fluoride
PEMA	: Polyethyl methacrylate	PBTP	: Polybutylene terephthalate
PBMA	: Polybutyl methacrylate	PFA	: copolymer perfluoralkoxy

[1] Very soft polymer; not applicable as structural material.
[2] Epoxy resins with different chemical compositions and different cross-link densities.

Appendix 7A

Table 7A: Moduli and fracture data

Polymers	4.2K				77K				150K		250K		290K		Ref.
	E' GPa	G' GPa	σ_{UT} MPa	ε %	E' GPa	G' GPa	σ_{UT} MPa	ε %	E' GPa	G' GPa	E' GPa	G' GPa	E' GPa	G' GPa	
HDPE	9.8	3.8	196	3.0	8.3	3.7	153	4.0	6.0	2.3	2.5	1.2	2.0	1.0	
POM	13.3	5.0			12.4	4.6			10.8	4.3	4.5	1.8	3.8	1.5	
PTFE	7.2	2.9	87	1.5	6.7	2.7	77	1.6	5.9	2.1	1.7	0.5	1.3	0.4	
PP	7.7	2.8			6.8	2.6			6.4	2.3		1.7		0.9	
PEEK	6.9	2.6	197	3.3	6.1	2.5	192	5.5		2.0		1.7	3.6	1.4	
PS	5.6	2.2	79	1.6	4.5	1.7	57	2.0	4.3	1.6	3.9	1.4	3.6	1.3	
PSU	5.2	1.8			4.3	1.5	130	7.0	3.8	1.2	3.1	1.0	2.9	0.9	
PC	5.6	2.2	177	3.3	4.9	1.9	156	6.0	3.8	1.5	2.6	1.0	2.5	1.0	
PMMA	8.2	2.9			8.0	2.7			7.5	2.6	7.2	2.1		1.7	1)
PI (Vespel)	5.8	2.1	157	3.0		1.9				1.7		1.4		1.3	
PI (Kerimid)		2.3				2.2				2.1		1.8		1.5	
PEI	5.9	1.7	166	3.4	4.9	1.5	157	5.2		1.5		1.3	3.0	1.2	
PAI	5.9	2.3	147	2.6	5.6	2.2	150	3.2		2.0		1.6		1.5	
EP I	7.1		150	2.4	6.1	3.1	150	3.1	6.2	2.5	4.2	1.6	3.1	1.3	
II	7.8	3.1	180	2.0	7.2	3.0	150	3.1							

Dynamic moduli E' at 10Hz and G' between 5–10Hz; Epoxies: EP (I): LY 556/Hy 917; EP (II): Cy 221/Hy 979 (Ciba Geigy)
Ref. 1) Perepechko, J. Low Temp. Prop. of Polymers, (1977), p.243, Pergamon Press

Appendix 8A

Table 8A: Permittivity ε_e'

Polymer	4.2 K	77 K	293 K	f	Ref.
PE			2.3	100 GHz	1)
	2.2	2.2	2.2	100 GHz	2)
			2.3	5 KHz	3)
			2.3	1 KHz	4)
		1.5	2.0	50 Hz	7)
PP			2.3	100 GHz	1)
	2.2	2.2	2.2	100 GHz	2)
			2.2	1 KHz	4)
	2.3	2.3	2.3	75 Hz	3)
PTFE			2.1	100 GHz	1)
			2.1	1 KHz	4)
	2.1	2.1	2.5	75 Hz	3)
PMMA			2.6	100 GHz	1)
			3.1	1 KHz	4)
PETP	2.6				5)
Nylon 11			3.3	1 KHz	4)
	2.4	2.4	3.2	2 Hz	3)
	2.4				5)
PS	2.5	2.5			6)
			2.4-2.7	1 KHz	4)
		2.5	3.0	50 Hz	7)
PC		2.0	3.0	1 KHz	4)
			3.0	50 Hz	7)
PVC		2.4	3.1-5.7	1 KHz	4)
			5.0	50 Hz	7)
Epoxy			4.1	60 Hz	4)
(My 740/D 230)		2.7	4.0	50 Hz	7)
(Cy 221/Hy 979)		2.4	4.6	50 Hz	7)

References for 8A

1) Afsar; M.N.; IEEE Transactions on Instrumentation and Measurement Vol. IM-36 No.2, (1987)

2) Hossam-Eldin AA ; Conference Record of the 1980 IEEE

3) Perepechko, I. I.; Low-temperature Properties of Polymers, p. 126; (1980),;Pergamon Press

4) Ku, C.C.; Liepins, R. ; Electrical Properties of Polymers, Chemical Principles, Hanser Publishers

5) Allan, R.N.; Dielectr. Mater. Meas.; Appl. Conf. 67-70; Inst. Electr. Eng. London (1970)

6) Mc Camman R.D. and Saba RG, Work RN, Journal of Polym Sci. A2; 7,10, p.1721; (1969)

7) Hartwig, G. and Schwarz, G. from: Advances in Cryogenic Engineering (Materials), Vol. 30; pp. 61-70; (1984); Plenum Press NY, London

Appendix 9A

Table 9A: Mechanical data

Material	Temp [K]	G_{IC} [kJ/m^2]	K_{IC} [MPa,m$^{1/2}$]	E' [GPa]	σ_{UT} [MPa]	ε_{UT} [%]
PS	4.2	0.16	1.0	5.6	80	1.5
	77.0	7.1[1]	6.1[1]	4.4	55	2.0
EP II	4.2	0.16	1.3	7.1	170	
	77.0		1.9[1]	6.1	150	1.9
PC	4.2	1.3	2.9	5.6	175	3.7
	77.0	4.8[1] 1.5[2]	5.2[1] 4.7[2]	4.8	155	6.6
HDPE	4.2	5.8	7.8	9.8	195	3.0
	77.0	9.8[1] 5.8[3]	9.6[1] 6.0[3]	8.0	155 153	4.0
PMMA	77.0	0.3[1]	2.2[1]	8.0		
PEEK	4.2			6.9	197	3.3
	77.0		3.4	6.1	190	5.2

[1] At LN_2 environment and a COD rate of $3 \cdot 10^{-5}$ m/s.

[2] At He environment and a COD rate of $5 \cdot 10^{-7}$ m/s.

[3] Stable crack propagation at a COD rate of $7 \cdot 10^{-6}$ m/s.

INDEX

Acoustic approximation, 26, 36, 38, 52
Adiabatic heating, 203-205
Angular frequency, 19
Arrhenius equation, 128, 133, 153
Asymptotic low-temperature behavior, 11
    loss-factor, 136
    moduli, 165
    permittivity, 184

Barrier jumping, 11, 12, 119
Basic properties, 4, 5, 10, 72
Binding forces, 5, 10, 141
    covalent forces, 5, 9, 10
    dipole forces, 9
    dispersion forces, 9
    force constants, 23, 24
    hydrogen bond, 10
    and moduli, 167
    rotational forces, 10 121
    structure of, 8
    Van der Waals, 5, 8, 9
Binding potentials, 9, 74

Boltzmann statistics, 22
Bose-Einstein statistics, 22, 44
Boundary effect, 13
    thermal conductivity, 105
Boundary scattering,
    thermal conductivity, 105-110
Brillouin zone, 20, 28, 32, 37

Calorimeter, 226
Chain structure, 6
Chemical composition, 6, 252
Chevron specimen, 198, 238
Compliance, 143, 148-150,
    tensor, 86
Configuration, 6
Crack propagation, 191, 194, 237
Crack arrest, 194
Crack tip zone, 191, 197
Crankshaft motion, 122
Craze, 197, 201
Creep behavior, 145

Cross-linkage, 8, 11
Cryogenic application, 3
Cryogenic measuring methods, 221
Cryostats, 222
    bath cryostat, 222
    evaporator cryostat, 223
Crystalline content, 11
Crystallites, 105, 107, 108

Damping spectra, 117
    polyacrylates, 125, 136
    theory, 127
Debye frequency, 42, 43
Debye function, 52-53
Debye law, 50-52
Debye temperature, 42, 43, 51, 53
Deformation behavior, 14, 141
    dependence on strain rate, 206
    elastic deformation, 14, 144
    energy elastic deformation, 142
    entropy elastic deformation, 142
    dependence on crystallinity, 155
    general survey, 141
    high load levels, 161
    viscoelastic deformation, 14, 144, 157
    viscous deformation, 14, 144, 159
Density of states, 22, 23, 36, 103
    helical structure, 41
    linear chain, 38
    normalization, 41
    and specific heat, 53
    three-dimensional isotropic solid, 39
        anisotropic polymers, 39
    variance and, 43
Dielectric properties, 175
    correlation with mechanical damping, 178
    damping, loss-factor, 176, 179
    high frequencies, 182, 183
    measurement, 230
    permittivity, 176, 177, 179
    polarization, 176
Dilatometer, 228
    inductive dilatometer, 229
    laser interference, 228

Dipole moment, 176, 184, 185
Dispersion relation, 25
　helical chain, 28
　isolated linear chain, 25, 27, 28
　measurement, 33
　and phonon velocity, 25
　three-dimensional array, 29-31
　zigzag chain, 27
Dominant phonon wavelength, 55, 104
　definition, 111
Double potential well, 12, 80, 128, 131
Dulong-Petit law, 50, 106

Electron
　deformation, 146, 177
　polarization, 177
Energy
　internal energy, 49, 106
　level, 130, 131
　zero point, 44
Energy release rate, 193, 199
　critical release rate, 210
Enthalpy, 49
Epoxy, 7, 58, 84, 112, 248
　group, 8

Fatigue cycling, 214-216
　self-heating, 212
Fatigue strength, 211
Fibers, 169
Fiber composites, 243
　fiber-matrix bond, 246
　lamination theory, 245
　mixing rules, 243
　processing, 249
　thermal stresses, 246
Force constant, 23, 24
Fracton, 31, 43, 103
Fracture behavior, 189
Fracture energy rate, 194
　critical energy rate, 210
Fracture mechanics, 192
Fracture models, 196
　Irwin, 196
　Dugdale, 197
Fracture stress, strain, 205
　dependence on strain rate, 206
Free volume, 12, 80, 168

Glass transition, 4, 5, 11
　main glass transition, 4, 5, 121
　nomenclature, 120
　secondary glass transition, 4, 5, 11, 121
　tertiary glass transition, 4, 5, 11, 121
Grueneisen parameter
　measurements, 87
　and pressure dependence, 85
　temperature dependence, 88
　tensor, 86
　thermoelastic effect, 88

Grueneisen relation, 70, 84
　anharmonicity of vibrations, 85
　macroscopic definition, 70
　microscopic definition, 71
　multi-mode, 85
　single-mode, 84
Harmonic oscillator, 21, 44, 51
Homopolymer, 6
Hysteresis, 164

Interchain forces, 5
Interchain interaction, 5, 6, 121
Interchain vibration, 69, 72
Intrachain interaction, 5, 6, 121
Intrachain vibration, 69, 72, 74

Liquid crystalline polymers, 170
Loss-factor, 127, 152, 176

Mean free phonon path, 100-105
Mechanical deformation, 141, see
　　Deformation
Mechanical testing, 234
Mises criterion, 203
Modulus, 143
　and binding potentials, 167
　complex modulus, 143, 152
　damping modulus, 143, 152
　definition, 142
　dynamic modulus, 151, 154
　initial modulus, 143, 158
　relaxation modulus, 143, 152
　secant modulus, 143, 158
　starting modulus, 148
Molar heat conversion, 63

Necking, 157, 203

Orientation function, 79
Orientation polarization, 177
Oriented polymers, 169
Oxygen bridge, 122

Permittivity, 176, 179
　asymptotic value, 184
　complex permittivity, 176
　and compliance, 177
　damping permittivity, 176
　definition, 176
Phase angle, 151
Phonons, 19
　acoustic phonons, 20, 34
　conversion, 31
　energy, 19
　momentum, 19
　optical phonons, 28, 32, 34, 41
　thermal phonons, 34
　velocity, 20, 25, 33-35, 101
Phonon wavelength, 20
　dominant wavelength, 111

INDEX

Place changes, 11-13, 121
    and damping spectra, 119
    and potential barriers, 121
Plateau
    damping, 134, 182
    thermal conductivity, 103
Poisson's ratio, 168-170
    definition, 168
Polymers
    amorphous, 7
    blend, 7
    duroplastic polymers, 7
    liquid crystalline polymers, 7
    semicrystalline polymers, 7
    structure, 6, 8
    thermoplastics, 7
Polymer matrix, 248, 250
Potential
    covalent, 74
    hydrogen bridge, 169
    Van der Waals, 74
Potential barrier, 12, 122, 162
    jumping, 12, 119, 128
Probability of states, 44, 50
    frequency and temperatures, 45

Relaxation
    compliance, 149
    parameter, 150
    single-relaxation, 145, 195
Relaxation time, 12, 127, 133, 152
    load assisted, 161, 162
    tunneling, 132, 133
Repetive units, 181, 252, 253

Scattering processes, 99-101
Self-heating, 196, 212
Shear expansion, 78
Shear modulus, 166
Shear fatigue, 216
Specific heat, 49, 101, 114
    definition, 49
    dependence on crystallinity, 58-63
    dependence on pressure, 62
    dependence on temperature, 54-58
    excess specific heat, 103
    theory, 50-54
Strain gauges, 236
Stress concentrations, 189
    microstress, 190
    molecular stress concentration, 190
Stress intensity factor, 192, 207-209
    critical factor, 207, 209
Superposition principle, 143, 158, 160

Tarasov-Baur approximation, 30, 40, 54
Tarasov theory, 40, 52, 60

Temperature gradient, 99
Temperature sensors, 225
Thermal conductivity, 99
    anisotropic conductivity, 113
    boundary scattering, 105-110
    and cross-linking, 112
    definition, 99
    general theory, 100
    plateau, 103, 113
    temperature dependence, 101
Thermal expansion, 69
    basic thermal expansion, 72
    and binding potentials, 74
    definition and survey, 69
    and temperature dependence, 88-91
Thermal expansion
    from bending vibrations, 75
    and cross-linking, 83
    from glass transitions, 80-83
    from stretching vibrations, 73
Thermal diffusivity, 114, 221
Thermal relaxation time, 114, 204, 221
Thermal resistance, 99, 106
Thermal stress, 190, 246
Torsion pendulum, 232
Transformation matrix, 246
    angle transformation, 246
Tresca criterion, 203
Tunneling
    parameter, 131
    plateau, 134
    relaxation tunneling, 130, 132, 182
    resonant tunneling, 130, 182
Tunneling processes, 4, 5, 12, 13, 120
Tunneling vibrations, 4, 5, 12, 13, 120

Unrelaxed state, 147

Vibrations, 15, 21
    bending (transverse), 21, 22-26
    interchain vibrations, 72
    intrachain vibrations, 72
    modes of, 15, 23
    multi-mode vibrations, 51
    polarization, 21, 23
    single-mode vibrations, 50
    stretching (longitudinal), 21, 22-25
    tunneling, 15, 136

Wagging, 123
Wave number, 19, 20
Wave vector, 19, 20

Yielding, 157, 201
Young's modulus, 166